THREE
QUITE ↓ very actual
CHEERS
for Worzel Wooface

Catherine Pickles   Hubble & Hattie

The Hubble & Hattie imprint was launched in 2009 and is named in memory of two very special Westie sisters owned by Veloce's proprietors. Since the first book, many more have been added to the list, all with the same underlying objective: to be of real benefit to the species they cover, at the same time promoting compassion, understanding and respect between all animals (including human ones!) All Hubble & Hattie publications offer ethical, high quality content and presentation, plus great value for money.

More great books from Hubble & Hattie –

WWW.HUBBLEANDHATTIE.COM

First published September 2017 by Veloce Publishing Limited, Veloce House, Parkway Farm Business Park, Middle Farm Way, Poundbury, Dorchest Dorset, DT1 3AR, England. Fax 01305 250479/email info@hubbleandhattie.com/web www.hubbleandhattie.com ISBN: 978-1-787110-58-8 UPC: 6-368 01058-4 ©Catherine Pickles & Veloce Publishing Ltd 2017. All rights reserved. With the exception of quoting brief passages for the purpose of revie no part of this publication may be recorded, reproduced or transmitted by any means, including photocopying, without the written permission of Velo Publishing Ltd. Throughout this book logos, model names and designations, etc, have been used for the purposes of identification, illustration a decoration. Such names are the property of the trademark holder as this is not an official publication.
Readers with ideas for books about animals, or animal-related topics, are invited to write to the editorial director of Veloce Publishing at the above addre British Library Cataloguing in Publication Data – A catalogue record for this book is available from the British Library. Typesetting, design and page ma up all by Veloce Publishing Ltd on Apple Mac. Printed in India by Replika Press.

CONTENTS

FOREWORD

Growing up in the north of England, Lurchers were already a recognisable breed for me. My Nan adopted a stray Lurcher – Rags – when I was still very young: our guardian angel who was kind of like a coarse grey mop on long legs. Rags shared our adventures and supervised play, only occasionally deserting us when he needed to patrol the fields surrounding the park for any intruding rabbits!

In later years, whilst working at the RSPCA in Blackpool, I came in contact with many Lurchers, very similar to our beloved Worzel Wooface. One in particular, Sally, was funny, naughty, and totally gorgeous. Like most Lurchers, she was quiet and cuddly at home, a complete couch potato, but totally unrecognisable as such once out in the fields. She had obviously been taught to hunt: standing stiff, nostrils scenting the air, scruffy golden hair blowing in the wind, she'd be off, leaving me dogless and frustrated ... fine times ...

My first ever dog was Penny, a Collie/Whippet cross, who was incredibly clever, could do all sorts of talented tricks, but who, like Sally would never come back when called. It's probably due to these two endearing dogs that I became involved in the world of dog training.

I so enjoyed Worzel's first two publications; they kind of took me back to the old TV series *Woof*, with Eric the dog, Eric, who was, in fact, Judy in real life, lived and worked with me for many years. The series portrayed her in a human-like role, and reading Worzel's adventures revived many fond memories

Luckily, Matisse (winner of Britian's Got Talent, and second left in the picture) and his furry friends, although clever, can't read, otherwise I fear they would start to rebel!

Hopefully, this series of books will shine a spotlight on the Lurcher, and people will recognise what wonderful, enchanting family pets they are, so that maybe others in rescue organisations will no longer be overlooked.

And now that we also have the first of Worzel's books for children (*Worzel says hello! Will you be my friend?* Hubble&Hattie), children of all ages can benefit from his wisdom, and enjoy learning about his adventures!

His cheeky and naughty escapades are sure to attract a great following of children and adults alike, especially if, at some time in their lives, they've been lucky enough to have also had a Worzel ...

Jules O'Dwyer
Winner, Britain's Got Talent 2015
Crufts International Canine Freestyle Champion 2013
20 years as Guide Dogs for the Blind, and Assistance Dogs professional trainer

INTRODUCTION
by Mike Pickles

I'm often asked what it is like to be 'Dad' to Worzel, and how I feel about having my life displayed to the world by him. I have to say, life as Dad is just as Worzel describes, although my love of ketchup is perhaps somewhat exaggerated. Life here is chaotic sometimes, and I find that I have to be 'the boring one' because all the other options have been taken. I'm sure many Dads will recognise this state of affairs in their own life.

We're not a remarkable family but we have one advantage over anybody else ... and that's Worzel. I think Worzel is a remarkable dog, and our love for him is unending. He got landed with us: he had almost no choice in the matter, and, for an anxious dog with issues, he seems to take things that happen here in his not-inconsiderable stride.

Worzel isn't brave but I've yet to see a dog who greets his fellow dogs with so much respect, for which we get admiring glances from other owners. We get less admiring looks when he decides to forget his recall and begins to resemble a fridge: a deaf, unplugged fridge.

So, whilst Worzel is a literary Lurcher and a superstar, to us, he is our dog, part of our family, and a constant source of joy and occasional bafflement. I'd like to say that we are equally baffling to him but, in truth, I think he's worked out exactly what each of us is like, and presents a fairly accurate view of us all. I do sometimes wish he wasn't *quite* so accurate: most of my work colleagues have read Worzel's books and are now convinced I do nothing but drink cups of tea, play on my computer, and "bog off down the boat."

That part, at least, is true.

DEDICATION

To Michael and Wendy Pickles – my wonderful parents-in-law – who Worzel loves just as much as I do.

With huge thanks to Alison Housden, who speaks 'Worzel' almost as well as he does now!

January

January 1 (early)

According to the fuge ginger boyman, it is very quite normal to wake up on Noo Near's Day with a pain in your head that feels like it is never going to end, and wondering why you is cuddling a traffic cone that seems to have cunningly become your New Best Friend. He also reckons it is quite very usual to not be able to remember wot happened the night before. The last fing I remember from last night is Mum and Dad getting hinvited to a Noo Near's Eve party, and Mum saying that as Worzel Wooface could not come, Dad must do taking me for a walk, and she would do her bestest best to make herself look booful-well-at-least-clean whilst Dad did making sure Worzel would be comfy and snoozy whilst they were out.

So, wot I want to actual know is why I can't remember anyfing else, and why *my* head does feel like it's been ripped off my boddedy and stucked back on the wrong way round. I does also want to know why Mum is hexhausted and stressed and tents, and also why she keeps lookering at me with a very quite concerned and wobbly red face. And why Dad isn't asleep next to me. Where he should be.

Somefing is quite very actual wrong. If my head didn't hurt as very much as it actual does, and if I wasn't so blinking tired, it would be my himportant work to try to find out. But I does fink for right very now, I is going back to sleep, and I'll do working it all out later.

January 1 (a bit later)

Sally-the-Vet just phoned. She does want to know how Worzel Wooface is feeling this morning. Mum's told her I are still very quite sleepy but not hungry. I does not want anyfing to eat currently but, according to Sally-the-Vet, that's fine and to be expected, and that doggys-wot-chase-things-and-then-fall-in-ditches-at-40-miles-an-hour and somehow-manage-to-stab-themselves-and-gouge-a-three-inch-hole-in-their-neck, and then need an Hemergency Hoperation on Noo Near's Eve, just as everyboddedy has gotted their party dresses on (heven Sally-the-Vet), probababbly don't want to eat their breakfast ...

Sounds quite very reasonababble to ... hang on a flipping, blinking quite very actual minute. SOMEBODDEDY NEEDS TO SAY THAT ALL ACTUAL AGAIN ... SLOWLY.

Mum says I went for a walk with Dad. I chased after somefing. I felled in a ditch and somehow managed to end up with a hole in my neck. As soon as Mum stucked me in the bath and started to wash off the mud and the god-knows-wot that was in the ditch, she did see a fuge, gougy, flappy bit of the Inside. Of. Worzel. Then she roared at Dad to heave me into the car, and we all raced down to see Sally, still half-soggy from the bath and half-muddy from the ditch, but now with a quite actual visibibble hinjury, and a Mum having hystericals.

Once Sally had dunned giving me some sleepy medicine, and did finish the job of cleaning me up, she saw that I managed to miss my jug-li-ar (wot is the himportant tube that carries all the blood from my boddedy to my head) by three actual millimetres. And I are very, very quite actual lucky to be an actual live Worzel Wooface, cos it is a miracle I didn't do dying. But I aren't dead cos I are still on the bed and I hasn't goned to the Rainbow Bridge, and Dad says I are going to be just fine in a few days. And if that is the case, I are going back to sleep again. Dad says he'll do joining me just as soon as he's peeled Mum off the ceiling where she's binned living since seven o'clock last night. I've had a good look and Mum isn't on the ceiling, but Dad says it's somefing called a metty-four. Like the Dog House.

Where he's binned living since last night as well. I are definitely going back to sleep …

January 2

Mum says I are a Total Twit. Dad says I should be grateful, even though he told me that a Total Twit is a fuge hidiot with knobs on. He also says I should not be hoffended but should probababbly feel quite very actual relieved and pleased that Mum has finally stopped shaking and making choky-blubbery noises, and at least she is actual speaking to me almost normally, rather than all that hello-fella-how-you-doing gentle talk wot is all very nice for a bit but is now making me feel a bit actual awkward, and like I want to get off the bed and not be the centre of attention any more.

I are doing much betterer than he is, Dad reckons. Mum hasn't founded any Suitable Words for Dad yet. He is still in the dog house and waiting for the fuge hexplosion from Mum about dusk and Sighthounds, and being a fortless hidiot wot doesn't concentrate or remember to do checkering there is nuffink for me to chase before lifting me over stiles and Letting. Go. Usually, he says, going back to work after Crispmas is boring and actual disappointing. And cold. But as the atmos-fear in our house is currently like ice, he reckons that even if he felled in the River Blyth, he'd be warmer. I hope he doesn't fall in the river to be quite actual honest. Mum's honly just managed to get all the mud and blood off the walls in the bathroom; I don't fink she wants to do it again …

January 4

Mum's found some Suitable Words for Dad. Tonight, she managed to say three ickle words. Before you do starting to himagine this was all soppy and lovey-dovey, I does fink I should tell you that the three ickle words were Six Fundred Quid. I wonder if Dad is feeling grateful and not hoffended now Mum is finally speaking to actual him. Like wot he tolded me to feel when she called me a Total Twit. Dad has tooked the news of my vet bill like the manly sailor wot he is: he looked at his moneys, realised that he isn't going to be able to afford the noo sails he wanted this year for his boat afterall, and has goned to sulk in the shed.

Tomorrow, Mum says, she'll remind Dad that because SHE isn't a Total Twit and SHE isn't a Fortless Hidiot, Worzel Wooface is hinsured to the hi-balls. But not tonight.

January 5

According to Sally-the-Vet, I are now well a-very-nuff to go for short lead walks so long as I do take it gently and am a sensible boykin. But I are not allowed no playtimes because my stitches are healing very nicely and she finks there is a good chance I won't have a scar. So I are not allowed to mess up all her clever and booful sewing by being a plonker with Merlin. Everyfing else about me, like bruising and pain and being sleepy from the hoppy-ration, has all goned. I feel perfickly actual fine. Well, I was until I heard about the lead walks and the no-friends, and definitely no bitey-facey instructions.

January 6

Lead walks, in actual general, is not somefing I are that keen on, but because of the hole-in-the-head fing, it's all I'm actual allowed to do. So, I are making the bestest best of it, and today I had a good sniff along the hedges and did my bestest practicing of cockering my leg and peeing. I aren't talented at cockering my leg, to be quite actual honest. Dad says it's somefing to do with my Centre of Grabititty; my legs is too long and my head is too heavy, so as soon as I lift one of my feets off the ground, I get all wobbly and feel like I'm going to fall over. I has looked and looked but I do not fink I have a Grabititty so I are quite very confuddled. It sounds very actual like wot happens to the fuge ginger boyman when he gets together with his friends. And cider.

    I hope I aren't going to start singing and howling like the fuge ginger boyman when he does drinkering cider. Last night, the fuge ginger boyman and his friend, who is called Who, began making up a folk song about Girls. It was all going quite actual well until they had troubles with their Grabititties, and then they staggered about everywhere. Mum tolded them to either shut up or go to the shed, and she wasn't-interested in their creative juices unless they wanted to get the mop and clear up all the cider they'd spilled on the floor. Mum reckons she can do without that kind of creative juice.

January 8

This morning, after my lead walk and my cockering my leg practice, and after Mum had dunned checkering that the fuge ginger boyman and Who hadn't dunned dying in the shed, or setting fire to it or spilling any more creative juices everywhere, we did decide to go and see Dad at the boatyard, so I could say Hello-I-are-a-luffly-boykin to Dad's friend, Keith. Keith is Dad's oldest friend, and Mum says they have been talking rubbish at each other for over 40 years. Mainly about boats. In quite actual fact, the ickle teeny-tiny hintricate hunimportant bits of boats.

    Now, it is a well known fact that I aren't wot you'd call a pushy dog. I don't do much in the way of Complaints to the Management but, to be quite actual honest, after about three hours of this hincredibibbly boring and complercated talk, I did decide that me and Mum had had a-very-nuff. So, I did howl. A proper fuge yowly howl just like Who and the fuge ginger boyman did last night with their cider and Grabititties, and singing-about-Girls. That did get Dad's attention, I can very tell you. He did stop talking and lifted me off the boat, and then he did watch me whilst I did exerlent standing on one leg peeing with my Grabititty. Wot after all the howling and singing, I has hobviously got.

January 10

There was a fuge kerfuffle in the kitchen last night, at 2 o'clock in the morning when Dad was the honly person still awake. He was playing a game on Uff-the-Confuser, and he did have his headphones on so did not hear the kerfuffle until Mum wobbled down the stairs, and poked him and asked him if he was aware we was being robber-dobbed. Wot he wasn't. He was too actual busy making sure his himaginary friends didn't die in his army game, and "gimme-a-sec."

Mum did a lot of hissy, hurgent hexplaining at Dad that there was someboddedy in our kitchen, and we was quite possibly all going to die In. Real. Life, wot, according to Mum's hurgently and hincreasingly hysterical hissing, is the life wot happens all around Dad when he has his headphones on. Most himportantly of all, though, could he go and look in the kitchen and scare away the robber-dobbers. Cos Mum didn't have a-very-nuff clothes on. Any clothes on ...

Once Dad had gotted over the shock of being poked In. Real. Life, and Mum standing naked in front of him, shivering and hissing about robber-dobbers, me and Dad did bravely and very quite heroically go into the kitchen. There was noboddedy there, apart from the cats, but none of them was actual asleep, and they did all look like the very worstest hexamples of themselves.

None of our cats is wot you'd call Normal. I aren't sure wot a Normal Cat looks like or acts like cos I does honly have hexperience of the five wot live here. And I are quite very actual sure that they can't be good hexamples of Normal Cats: if all cats were like our cats, cats would be hextinct.

Gipsy, who is the senior cat and In Charge, was standing on the table, puffed up and trying to be hassertive and haggressive, but also having a coughing fit cos, at the end of last year, she did develop assma. Frank was trying to squeeze through the cat flap which is a fuge job for him anyway cos he is the size of a sheep: the honly bit of him I could see was his tail, and *that* was all puffed-up and spiky, so I do fink the rest of him was probababbly the same. I do suspect he was well and truly stucked, and honly going to get out of the cat flap once he did calming down and stopped being puffed-up and spiky.

Mabel was having hystericals trying to shove Frank through the cat flap so she could get through it, and run away and hide in the shed for HEVER, but also attempting to avoid Gandhi who fort Mabel's meltdown was a fabumazing excuse for a game, wot she Did. Not. Happreciate. Mouse, being slightly very a lot fick, was wandering around looking confuddled until she saw Dad, and then she attacked him with her claws and tried to climb on his head.

You woulda fort that me and Dad arriving in the kitchen ready to Save The Day and get rid of the robber-dobbers would have calmed down the cats. It didn't. For a start there was no sign of any robber-dobbers, so I did all the helpful fings I could fink of: I did helpful shoving of Frank out of the cat flap by pouncing on his tail, broked up the fight between Gandhi and Mabel, bravely and wise havoided Gipsy cos she had stopped coughing and was doing turning into a psycho ninja murderess, and then danced around under Dad's feet, wondering how to get Mouse off his head, cos she had imbedded most of her claws in his forehead. Every time I woofed, she just clinged on tighter.

By the time me and Dad gotted back to the hoffice, Mum was hiding hunderneath the desk wearing nuffink but Dad's headphones, trying to hexplain to his himaginary friends that Dad was dealing with robber-dobbers in the kitchen. Which is why they'd all dunned dying in the game. And cos Mum was not completely very sure if there was a webcam on Dad's confuser, and did fuge panicking about whether they could see her as well as hear her, she'd decided to hide hunder the desk just in case.

Once Dad had dunned exerlent convincing Mum that he didn't have a webcam and she could come out from hunder the desk and go back to bed, all of Dad's himaginary friends did want to know if everyfing was actual okay at our house. The last fing I hearded before I went back to sleep was Dad using his army game proper talk and telling them that everyfing was 'Situation Normal.' Well, *our* version of Normal, anyway ...

January 11

We very actual *was* robber-dobbed last night, but not by a hooman robber-dobber. It was a burgling cat, and when Mum went into the kitchen this morning, he was back again, sitting on our kitchen table and making himself quite very actual at home, helping himself to *our* cats' food. And for some actual unknown reason, all of our cats were letting this happen. Heven Gipsy. I do fink they was probababbly shocked because the Burgling Cat do look like a cross between Frank and Mabel, wot is a-very-nuff to confuddle anyboddedy. I did decide I was not shocked or confuddled, but mostly that this was hunacceptababble. So I chased him out of the cat flap.

January 12

Dear Burgling Cat

I are sorry I has just chased you out of the cat flap again. I did not mean to do giving you a art attack. In very actual fact, I did do you lots of favours.

1 You don't live here. I know cats can be a bit variable about where they do live, so maybe you did need remindering

2 You is not a skinny, half-starved cat. You is already fuge, and you does not need to be getting any fuger or you won't be able to fit back in through your own cat flap

3 Dad says we already have five cats and we don't need no more. Also, the last time a cat did trying to move in here, one moved in and then five more appeared! I are quite actual sure you don't want that happening to you ... especially as you are a boy

From your luffly boykin

Worzel Wooface

Pee-Ess: You seem to be missing a tail? Please tell me this wasn't me. I can't find it here anywhere and I'm having a bit of a panic now ...

January 14

My Mum has hadded a word with the burgling cat with no tail and has discovered that, first of-very-all, a) I are not responsibibble for the missing tail, and b) that the burgling cat with no tail is quite very actual cheeky, and could-we-please-chuck-him-out-of-our-house-before-it-becomes-a-nabbit-and-he-tries-to-move-in! I has decided that I will be In Charge of this himportant work, cos if Frank and Gipsy-the-Cat get very hinvolved, fings could get hexpensive. I has also founded out that the burgling cat with no tail's name is Grey Cat. Now, I doesn't want to cast Nasturtiums on other people's name-choosing ability, but I do fink if you are going to give your cat a describing name, then I fink you should try to mention the himportant and not usual fings. Like the fact that he's got No Tail. Cos there is lots of grey cats in the world but not that actual many with No Tail.

January 15

It turns out that the burgling-cat-with-no-tail didn't start off life with no tail, though he did start off being grey, which is why he is called Grey Cat. Then he losted his tail, but it was too actual late to do changing his name, even if peoples do fink that his Mum is a bit actual bonkers. Or missing the himportant point ...

I does know hexactly how she does feel. The hoomans wot live here all have their own names, but I do mostly fink that *my* names for them is much betterer, apart from the previously ginger one. This is the name I choosed for the smallest hooman wot does live with me, and it was very perfick cos she used to actual change her hair colour all the blinkin' time. And then, according to Mum, she did grow out of that redickerless hair-changing-stuff. and it has binned proper ginger for a long time now. But she is stucked with the name previously ginger one cos once you've got a name, you can't just do changing it or everyboddedy will get confuddled. Mum reckons that there is about as much chance of the previously ginger one sticking with being ginger as there is of the cat-with-no-tail hunexpectedly growing a new tail. It's binned ginger for six actual months now so I are sure she will be getting bored soon, and my name for her will make sense again. Mum says she's hoping I are wrong ...

Everyboddedy does know when I talk about Mum that I are talking about a female hooman who is actual quite stressed and tents most of the time, but also In. Charge. Whether she likes it or actual not. And everyboddedy who is a Dad knows that it is their himportant actual work to do stopping the Mum having hysterical and hiding in the shed if it looks like there is any Going Out or Dressing Up to be dunned.

The fuge ginger boyman is fuge. And ginger. But he really, wheely doesn't fink it is quite actual fair for me to still be calling him a boyman. He's twenty-one now, and finks he's a Man-Man. From the giggling and cuppateas spluttering wot came from Mum and Dad when he did suggest this, I fink I are quite correct and safe to keep calling him a boyman. Mum reckons if the fuge ginger boyman can go more than a week without losing his wallet, or spilling cider everywhere, or needing to borrow some moneys cos his rent is due and his laptop is dead and he's going to Fail. His. Degree if Mum doesn't do somefing to help, then he might be a Man-Man. Dad says being a Man-Man is not nearly as good as wot the fuge ginger boyman finks; he'd love to be a boyman again with no worries or troubles or famberly ...

Or dinner, Mum says.

Dad says he's going down to the harbour to play with his boat and remember his Lost Yoof. The fuge ginger boyman has gone with him. To see if his lost wallet is down there, too.

January 17

Happy Crispmas everyboddedy! Sort of. Tonight, all my famberly has goned over to visit Gran-the-Dog-Hexpert to have Crispmas Day. If you is finking that my famberly is one of those clever ones wot do Crispmas in January so they can take fuge hadvantage of all the shops that have half-price presents and exerlent foody bargains, I are sorry to say that it is not cos of that fing, but because they is all quite very disorganised, and did have actual proper Crispmas Day in Corny Wall. And then I tried to chopper off my head. And Gran-the-Dog-Hexpert was actual busy. So they is doing it all tonight.

I aren't allowed to actual go with them, wot you might fink is a shame, but I are still not allowed to do playing, and I has to say I are quite very relieved I can stay at home. When Gran is on her own with her doggies, they is all very quite well-behaved and actual peaceful, but when they has hooman comp-knee, they do all go completely very actual bonkers and do a lot and a lot of showing off. Mum reckons I would find it all a bit too much. Dad finks he will find it all a bit too much as-very-well, and tried to tell Mum he was staying behind with me cos he doesn't want to get Cavved By The Cavs.

Cavved By The Cavs is wot happens when you do try to sit on the sofa in Gran-the-Dog-Hexpert's house, and all of Gran-the-Dog-Hexpert's Cavalier King Charles Spaniels do use their cunning and vicious shark-like skills to savagely and hagressively cuddle their prey to Very. Actual. Death. I has losted count of how many Cavs there are wot live with Gran now, but it is at least a billion. Or that's wot Dad reckons, at least, which it why he wants to stay home with me.

First of-very-all, the Cavs do pickering their victim: always the person who they does know the least well. Then, they do all try to sit on their victim at the same time. and have a bit of a hargument about whose turn it is to be closest. And who is in charge of licking the victim's hearholes, and who is the bestest at crawling under jumpers. I fink there is a special prize for the doggy wot finds the hooman's belly button first. Has you ever seen one of those pictures of peoples standing in Trafalgar Square pretending to be a tree whilst pigeons do using them as a tree? And do poos on their head? It's like that, honly with Cavs. But no poo, fortunately.

Heventually, the victim does manage to spit out the Cavs' tails for long a-very-nuff to yell at Gran-the-Dog-Hexpert to come and rescue them. And when she walks in it's like the quickest and mostest himpressive game of musical bumps you has ever seen, cos all the Cavs fall to their bellies and do lookering at her with hinnocent eyes, and cute ickle tail wags. And then Gran-the-Dog-Hexpert wonders wot all the fuss is about and tells Dad to Stop. Being. So. Redickerless.

This time, though, I has got to say that Gran-the-Dog-Hexpert might have to do believing Dad, cos when she walked into the room, both Mum and the previously ginger boyman were standing on the table laughing and taking photos. And exerlently avoiding being Cavved By The Cavs.

January 19

Today is my hunofficial 'Gotcha Day.' Dad finks it's in a couple of weeks' time so don't say anyfing to him! Two years ago today I harrived in Suffolk. Mum says she knew from the second I lummoxed through the door that I wouldn't be going anywhere. Then me and Mum did epic helping-Dad-fink-it-was-his-idea and showing him that I was exerlent for chatting up girls down at the harbour. and making Mum's bum get smallerer.

It didn't take Dad that long to realise that I was fabumazing, and that Mum was right. When Mum has a Nagenda, Dad generally finds himself agreeing with her quite actual quickly. Without realising wot's going on. Happarently, it's cos she's a female hooman, and they always get their own way. Heventually. So you might as well save yourself the 'talks,' and just say 'yes dear' straightaway. So, me and Mum are having a sneaky not-celly-bration, and later on we're going for a walk and coffee and cake and a bitta sausage. Just don't tell Dad ...

January 20

I are getting quite very actual bored of lead walks now. My head might be needing a rest but my legs want to do running and fundering along; my chest needs to get puffed out and my tail hurgently needs to practice its whipping hacrobatics for keeping me upright when I do sudden turning. And I does really, wheely miss my playtime with Lola and Merlin, and my noo friends across the road.

Mum has decided that I need hentertaining, and I aren't sure I'm that very pleased about this, neither. It's a quite very well knowed fact that I aren't keen on hobedience or trick training or toys, or any of those quite actual normal fings wot most doggys do like. I can get quite actual anxious and worried, and in actual general, when any of those fings is suggested, I do fuge over-finking ... and then run away and hide until Mum forgets her stoopid ideas.

Today, though, my not-doing-running-and-playing-with-my-friends, and being a bit actual full-a-beans did mean I forgotted about being worried. Mum was quite actual smart and kind when she boughted me a squeaky chicken toy, and I has had a quite exerlent time making it squeak, then jumping on the sofa and spinning round and round until it do squeak again.

It was all going quite very fabumazingly well, until Mum got too very fusey-fastic watching me, and said I was a clever boykin in a loud and hexcitababble voice, wot made me frighterened, and I skulked off to hide under the table. I was having a Private Moment of Madness. And I didn't want a haudience, and noboddedy joining in.

Mum's muttering in the kitchen now about overfacing-Worzel-and-will-she-ever-learn. So whilst she was busy doing making a cuppatea and remembering very hard not to crash fings about, even though she was cross with herself and very wanted to, I sneaked out from hunder the table and gotted my toy. And gived it a little squeak to let her know I had recovered from her hinterfering. And now she's trapped in the kitchen cos she doesn't want to hupset me again.

SORRY ABOUT THE WET PATCH

Sorry about the Wet Patch
I fink it might have been me
I got a squeaky toy today
Wot made me do a wee

I didn't mean to do a wee
I blame it on the toy
It made me all hexcited
Like a baby puppy boy

Sorry about your soggy foot
From standing in the wee
At least the carpet's dryer now
There's no wee left to see

January 21

Dad has only quite very actual forgotted Mum's birfday. For some reason the previously ginger one has not reminded him like she does usually, because she's not been able to get out of her bed for the past few days. When she gets like actual this, very actual quickly her bedroom turns into a hexplosion of cups and cola bottles, and noboddedy can actual get in there without breaking themselves. Or somefing himportant that's slidded off the bed and onto the floor, so I don't fink Dad has binned able to communercate with her.

Mum's not that hoffended that Dad has forgotted her birfday. She says she has had billions of birfdays and they all mainly go the same way, so Dad forgetting has made this one different and memorable, and she's mainly hintrigued about whether he will actual remember before the end of the day. But, more himportantly, she's wondering if her birfday will hencourage the previously ginger one to get out of bed and actual eat somefing other than crisps. That would be the bestest birfday present she can fink of, currently.

January 25

Sally-the-Vet says I are hofficially betterer, and can do having off-lead runs and playtimes again. My scar is fabumazing! It's completely hinvisible and, to be quite actual honest, unless you knew wot had happened you would never be able to tell from lookering at me.

Hunfortunately, Mum does know hexactly wot happened to me, and she's feeling quite very anxious and nervous about letting me off my lead. Hespecially, if there is anyfing like a ditch or a hedge or a fence that I could run into. She used to save up all her hystericals for cars, but now she says she's scared of letting me off my lead anywhere: she's lost all her confitents and doesn't know wot to actual do. Dad keeps remindering her that it was a freak-accident-and-maybe-I'll-be-more-cautious-now, and if she's worried, does she want Dad to take out Worzel?

Sometimes, I do fink my Dad is too brave for his own actual good, and I fink he is now actual regretting trying to be hencouraging and helpful. Mum says she needs his advices about walking Worzel like a hole in the head. Another one. To match the one Worzel got.

January 26

Mum is being very hunfair I do fink. She can't work out where she feels safe walking me, but she won't let Dad walk me. Meantime, I are going quite actual bonkers here getting bored.

****FORTS ON GETTING HURT AND GETTING BETTERER****

- ❖ I has got no idea wot happened
- ❖ The hole very hidea of me learning not to do it again is losted on me. Cos I has got no idea wot happened
- ❖ If it don't hurt, I aren't hurt. Recooperating isn't somefing I fink I need to do
- ❖ I got 'over the shock' about ten minutes after it did happen. I does not hunderstand why you is still not 'over the shock' two weeks later. It didn't happen to actual you
- ❖ I can't 'take it easy': I are a lurcher. I do stop and go. Pootling along is not in my jeans. If you does want to pootle, borrow one of Gran-the-Dog-Hexpert's cute-but-very-senior Cavs
- ❖ Anyfing you does let me do when I is hinjured is okay for me to do when I is betterer
- ❖ I can find a small patch of grass in a fuge field wot a strange doggy did pee on three days ago
- ❖ Finding a tablet (and actual hignoring it and not eating it) in a small bowl is really, wheely not that hard
- ❖ If there are a choice between having a scar and seeing my friends, I'll see my friends, fanking oo kindly
- ❖ Even though my boddedy is a bit broken, my brain actual isn't. All the energy I aren't using running around and remembering my manners with my friends will get used up on actual somefing. Like deading cushions and eating letters and chasing our cats out of the catflap.
- ❖ Sorry about that fing
- ❖ Please, please, quite very actual flipping, blinking *please*, can we go for a proper playtime now?

January 28

At flipping, blinking last! Mum's brain started to work again today. I do fink it is quite actual possibibble that me eating a quite very himportant letter might have had somefing to do with it, though. I fink it did focussing her actual mind on the himportant fings in life, like taking me for a walk, where I can do running without her worrying about me bumping into fings.

Mum's remembered we have a beach. I has got no very-a-tall hidea how she managed to forget this fing cos it's quite actual fuge, and not hexactly wot you'd call hidden away. And it's fabumazingly perfick for a luffly boykin who needs to run and have a blast about without worrying about cars. Or ditches. There is no ditches at the beach, but there is some friendly doggies wot, if I are a respectful and gentle boykin, will play with me. There is also waves for lying down in when I did get too hot ,even if this did Orrify everyboddedy, wot with it being January and cold, and lots and lots of super smelly stuff had binned washed up with the tide.

Mum says she is feeling quite actual much betterer now she has remembered where there is somewhere safe and hinteresting for me to walk, but I has to say, if she's feeling betterer, it is actual nuffink to how much betterer *I* do feel.

January 30

Southwold beach is perfick for luffly boykins like me, hespecially in winter, when there are no small people making piles of sand wot is-a-castle-and-you've-trodded-on-it-you-oaf. The sea is all quite very angry and crashy, and I can do zoomie running around and fortful checking that the sea still tastes disgustering. It is hexciting and frilling, and very sometimes, another doggy do want to play with me. But even if there is noboddedy there, wot in January is very actual normal, there is still plenty for me to find hinteresting, and I do not have to fink about bogging off.

Bogging off on Southwold Beach is quite actual hard to do, as it goes on and on for HEVER, and honly stops once you get to the harbour entrance, wot is actual quite scary and off-putting, and doesn't look at all a very good place for swimming. In the other direction, the beach stops in Scotland, I do fink. I hasn't been that far along the beach yet, cos I generally get puffed out by the time I get to the Peer.

A Peer is a fuge, long, sticky-out bit of wood and pavement and shops wot does go over the sea. It's a bit like a bridge wot people forgotted to keep building. Either that, or they did realise that they would have to keep going to actual Habroad, which is blinking miles away, before they gotted to put down the other end of the bridge. So I fink they builded a bit of it, then realised they had made a fuge mistake and just stopped. And pretended they was building somefing different in case other peoples did call them stoopid for not realising. So they called it a Peer and said peoples could walk along it and Peer into the water and Peer at the boats sailing along.

The bestest fing about the Peer isn't the sticky-out in the water end. At the other end, the end that's still attached to the land and not trying to get habroad – there is a cafe, wot we did visit today. Our visit did not start off in the bestest way cos Mum did remembering about sit again, just when I fort she'd finally forgotted about the blinking fing. Which is quite actual annoying and hinconvenient when you aren't the right shape for sitting, and in quite very general would actual rather not. Today, she decided to remember about sit in a stonking great sand-blasting hurricane. I forted about doing Complaints to the Management and hobjectoring and all sorts of fings that aren't sit, but Mum got all hinsisterent and said fings like we-are-not-going-for-a-cuppatea-until-you-do-sit, so I did it flipping blinking quick cos I really, wheely wanted to get out of the wind and the sand and the cold. And also do going for a cuppatea cos in very general, where there is a cuppatea, there is very hoften a bitta bacon or even a sausage. I aren't stoopid; I has binned going to these cuppatea places for quite a actual long time now.

Only trouble, is, now Mum finks I are all fusey-tastic about doing sit. And even that I like it, which is redickerless. Hobviously. I just wanted to get it over and dunned with so we could get to the cafe for a bit of warmth. And a sausage.

January 31

On sunny days in winter, if it is the weekend, billions of peoples visit the beach and walk along somefing called the Prom-and-Hard. A Prom-and-Hard is a posh name for a concrete path where hoomans can pretend they are at the seaside and hignore the sand and the sea. Dad says in olden days peoples used to put on their bestest clothes and walk along the Prom-and-Hard so that they could Be Seen, and other peoples would think they was posh and himportant. It does sound hexactly like a Dog Show, but for peoples, and I does fink it still actual happens now at the Prom-and-Hard.

In the Dog Show for Peoples you do get lots of points for wearing shoes wot have high heels and red bottoms. There is also points for wearing a large, furry, white hat and sunglasses at the same actual time. But the mostest points come from having a tiny person in a fuge trolley. In very general, these either have to have three wheels with fat tyres that look like they've been pinched from a lorry, or they has got to have four thin, very tall wheels and an ignormous handle like a lawnmower. Little trolleys, that can actual be pushed along without taking up the hole of the Prom-and-Hard wot are nippy are actual frowned upon. And you cannot win the Dog Show for Peoples with them.

There is also a Nagility compertishon for anyboddedy who manages to totter along on their high heels, keep smiling and pretendering they is having a fabumazing time, while they does juggle a fuge trolley with a tiny person, a takeaway coffee cup, a toddling small person and a furious pug who is actual quite desperate to go on the beach and come in the sea with me. But it can't cos the trolley will sink in the sand and the high heels costs fundreds of pounds, and the toddling small person might get sandy or even Wet. Wot is the worstest fing in the hole wide world that can happen, and will himmediately get you chucked out of the Dog Show for Peoples.

Today, there was lots of people Prom-and-Harding, but instead of lookering at each actual other, they was all watching me and my doggy friends having a fabumazing time whizzing in and out of the sea and roaring along doing bitey-facey.

Apparently, us doggies are very hentertaining, and most peoples fink we be quite actual bonkers paddling in the sea and tasting the seawater, and don't-we-know-it's-January? I fink they must know that the water always tastes disgustering, cos they all stay up on that concrete shelf fing and laugh at me checking again. And again. And again.

Visit Hubble and Hattie on the web: www.hubbleandhattie.com
www.hubbleandhattie.blogspot.co.uk • Details of all books • Special offers
• Newsletter • New book news

FEBRUARY

February 1

Yesterday was my Hofficial 'Gotcha Day,' and everyboddedy did forget, wot is hunacceptababble, I do fink. A Gotcha Day is the special day when a rescue doggy does become part of a famberly, and it is usually celly-brated, or at least remembered. To be quite actual honest, my Gotcha Day is not that sig-niffy-cant to me because I are quite very not good at telling the date or the time, and I do not like fuge fusses, so parties are not my favouritist fing. But there is usually some special words and some good pretendering that Dad did have a very small hinvolvement in the decidering, and that Mum did at least do a little actual bit of sensibibble finking, rather than just falling in quite very actual love with me the moment I arrived.

I aren't hexactly sure why my Gotcha Day did get forgot, but whilst the hoomans was out, I did exerlent showing peoples why having a dog is a Fuge Responsibility and how I does not actual happreciate being forgotted about. And peoples all rushing out of the house first fing without having time to take me for a walk is not actual making me a pri-orry-tree. So, I did have my own actual quite fabumazing celly-bration ... with the previously ginger one's new boots.

When Mum came home and saw bits of boot all over the bed and the landing, she did a lot of telling herself off about taking-Worzel-for-granted, and heven though Worzel is now all-growed-up-nearly-three-years-old, he is still a dog. And dogs cannot be hignored or not tooked for a walk, heven if everyboddedy has stuff to do.

When Dad came home, he did a lot of telling the previously ginger one not to whine when he has tolded her a billion times about putting her fings away. The previously ginger one said a quite very actual lot of fings but none of them is suitable for plite comp-knee.

I aren't saying nuffink, cos actions speak louderer than words. This morning, I has binned for a fabumazing walk down at the beach, and now I are snoozing. And not eating boots.

February 3

Mum is in fuge troubles. This weekend she has binned lookering after Gran's dogs at her house, and she has broked Gran's confuser. On a scale of breaking stuff at Gran's house, the confuser does come a quite very long way down the list of Fings You Must Not Break, wot starts with all of Gran's dogs and her cats, then her freezer and her gates, but it's still broked. Mum isn't happy and she does want Dad to fix it.

Dad is quite very actual good at fixing confusers but honly, he reckons, if Mum hasn't tried to fix them actual first. Cos then he's got to undo all Mum's fixing, wot has no logic but just a lot of confuddled button pressing. Apparently, it isn't the inside brainy bit of the confuser that Mum has broked but the outside

Alfie-bet pressing bit. One of the keys was quite actual sticky so Mum tried to clean it ... and it just pinged off. Then she saw a lot more gunky stuff, and tried to clean *that,* and more letters pinged off. She tried to stick 'em back on but they had clips, and it is fiddly, and ... Now there's honly three letters lefted attached to the keyboard fing, and please could Dad come and fix it? And also if he has a spare letter J could he bring that, too, cos she's got a terribibble feeling one of the Cavs has eated it.

February 4
Mum's founded the letter J. She isn't that actual keen to tell Dad where she founded it, so Dad is refusering point very actual blank to fix it, and says Mum will have to do confessering her stoopidness to Gran. Or quickly go out and buy her another keyboard. Wot might be easier and a quite actual lot less noisy and hab-dab-making than trying to tell Gran-the-Dog-Hexpert where the letter J has binned for the past 24 hours ...

February 5
Last year, me and Mum spended a long actual time trying to get Mabel to believe I was not going to eat her, and although it tooked for Hever, she will now come into the kitchen and eat her dinner, and not run away every time she does actual see me. So, you would fink the hole fing was sorted now, and Mabel could get on with being a nearly-normal cat (if such a very fing hexists), and stop worrying.

Mum was quite actual pleased with our progress, and did fink she had dunned a good actual job. But now, Mabel has dunned deciding that if she is not allowed to be scared of me, she's going to be scared of the oven fing that heats up cuppateas wot have binned forgotted about. Dad says the microwave has been sat in the same place for the past ten years, and hasn't growed wings or started singing the Natural Hanthem, so he doesn't hunderstand why Mabel has suddenly decided it is a monster. But she won't go near it, and now noboddedy can heat up their cuppateas when Mabel is in the kitchen cos she has hystericals and charges out of the cat flap like she's binned shot.

Dad's dunned a lot and a lot of checkering the microwave to see if it is secretly sending out sparks or zapping Mabel without the rest of us knowing about it, but there is nuffink wrong with it so he doesn't hunderstand wot Mabel's blinking problem is.

I do. Mabel is a scaredy-cat. Mabel is one of them cats wot has got to be scared of somefing. And now she's not scared of actual me anymore, and cos she is a cat wot doesn't have a lot of himagination, she just looked around for somefing else to be terry-fried about, and decided that the microwave would have to actual do!

February 7
It would be a quite very nice fort to believe that Gandhi, our youngest cat, was named after a fame-mouse Man-of-Peace wot changed the world. He wasn't. His full name is Gandalf-Gandhi-for-short-Terrance-III, and he was

named after a committee of teenagers drunked a bottle of vodka. (And as far as I can remember, it wasn't that peaceful, it was quite very actual noisy, and noboddedy could agree wot he should be called, which is why he has such a redickerless name.)

And there is no blinking way anyboddedy would fink Gandhi is a Cat-of-Peace at the moment, cos he's binned fighting. He is trying very quite actual hard to be in The Nile about this fact, but he is failing. Dismally. The previously ginger one is convinced that he is still her soft squishy very quite fusey-tastic hooman-loving Onorary Lurcher, but I fink she's in The Nile with Gandhi as-very-well.

Mum says as soon as she's founded the tweezers and removed the huge claw that is himbedded in Gandhi's nose, the previously ginger one can have her ickle babykins back. Until then, he's a blinking-great-thug, and could he please stop squabbling with the neighbours' cats and start behaving a bit more like his namesake, before fings get hexpensive.

February 8

Every-so-very-often a bit of Dad's leg wot is called a Nee does decide it doesn't want to, well, be a bit of Dad's leg anymore. It just wants to lie on the ground hurting and sulking, and does refusering to carry Dad around any-actual-more. Hunfortunately, Dad's Nee isn't that keen on giving him any warning about not-being-a-bit-of-Dad's-leg-today, and it clapses. And then it does some more clapsing and hurts for about a month, and then heventually, it does decide to work again and stop being so redickerless.

Dad says he doesn't need to go to see the Gee-Pee, but he does need to moan about it a lot and a very lot, and he can't do anyfing apart from play on Uff-the-Confuser and ask for cuppateas. And visit his boat every day. But he can't empty the bins or help Mum in the garden or stack the dishwasher.

I are quite very shocked about this: I didn't actual know Dad knew how to stack the dishwasher, with or without his sulky, clapsing Nee.

The biggest actual problem for Dad, apart from the dishwasher, or Mum trying to get him to go to the doctor, is the stairs. Dad says it is my very, very hurgent and himportant work to keep-out-of-the-blinking-way on the stairs in case his leg clapses hunexpectedly. Dad's tried to suggest I go first on the stairs, but this is not wot we has practiced and hagreed for the past two years, and so I aren't doing it. And overtaking on the stairs is not my favourite fing to do, ever since I brushed past Mum and knocked all the clothes out of her hands, and ended up wearing several jumpers and a pair of knickers wot was frightening, and didn't need no photos takering by the previously ginger one, even if I did look cute-and-hilarious. It was not good for my digger-nitty. A-very-tall.

Trouble is, Dad is taking a very long actual time going up the stairs at the moment, and I keep forgetting about the poorly Nee, and finking he has founded somefing hinteresting on the stairs that I should do lookering at or being helpful with. And when we is coming downstairs together after our luffly morning cuddles and cuppateas, I has got used to a certain actual speed of trottering. But then he stops, and I go walloping into the back of him all hunexpectedly.

Soon, Dad reckons, one of us is going to Come. A. Cropper. Dad's Nee is going to give way and I are going to get squashed by a fuge lot of jumpers and knickers and trousers, with Dad still inside them, wot will hurt a lot more than my digger-nitty.

february 10

I has founded out somefing Horrendous. And Hawful! Today, when I was at the beach, I did get to have a playtime with Levi, who is a super-duper Saluki. Levi can run almost as fast as me but that is very okay because he does also have A Recall. So he is wot is known as a good hinfluence.

But while me and Levi was having a run around, Levi's Mum tolded my Mum that she'd hearded from Bess-the-Lurcher's Dad that the Cow-Sell has got a plan to stop dogs playing on Southwold Beach. Our beach. MY BEACH.

The Cow-Sell, Mum says, is like the poxy Guv'Ment only smaller and honly hinterested in our bit of Ingerland. And it should know perfickly well about the people and the businesses and the houses in our area, and wot matters to our bit of Ingerland. But happarently, they does not, cos the Himportant and Stoopid Cow-Sellers on the Cow-Sell wants to stop dogs going on the beach, not honly during the summer, when there is lots of small peoples about and wot is very quite fairy-nuff, but during the winter as well. Doggies will still be allowed on the Prom-and-Hard on leads but they won't be allowed on the beach.

As far as I are concerned, this is very, completely redickerless, and means that all the doggies are going to have to do walking along the Prom-and-Hard with the buggies and the wheelchairs. And the shoes wot are scared of the sand. If the shoes are scared of the sand and the sea, I do actual Dread. To. Fink wot they is going to make of a fuge and wet and soggy Worzel Wooface wriggling up and down their Prom-and-Hard, covered in sand and sea and God Knows Wot. The honly peoples who want to be on the beach are the peoples with the dogs, and they'll be the honly ones who won't actual be allowed!

Anyway, according to Levi's Mum, Bess-the-Lurcher's Dad has told Ninja-the-Terry-fried-Terrorist's Mum, and she is going to Do. Somefing. About. It. She has started a Pee-Tishon which is when lots of peoples sign a bit of paper, and when it gets heavy enough she will go and bonk the Himportant and Stoopid Cow-Sellers over the head with it. I might have gotted that bit wrong to be quite actual honest but Mum was yelling at Dad about it, and everyfing gotted quite actual loud so I stopped listening.

february 11

Actual sometimes, Mum is Not Nice about Dad. Today, she did tell the Gee-Pee's lady wot answers the phone that Dad's name was Mud, and that he might be getting a sig-niffy-cant head hinjury quite actual soon if the Gee-Pee doesn't Do Somefing About His Somefing-beginning-with-B-Knee. I fink this is quite very actual hexcessive; emptying the dishwasher isn't that actual himportant. Is it?

February 12

This beach fing is really, wheely starting to worry me now. After all the dramaticals at the beginning of the year, and me trying to chop off my head, I has really binned henjoying the beach. And Mum has felt safe letting me off my lead in a place where there is no cars or fings for me to chase, so she has binned much actual happier as well.

But now with the Cow-Sellers Beach Ban plans, Mum is getting very quite anxious and tents, and trying very hard not to hinterfere. Usually, mostly, very quite sometimes, Mum hinterfering can be a good fing, and then, haccording to Dad, she ends up In Charge and he has to cook his own dinner too actual often. But in this case, I don't fink Mum being In Charge will help a-very-tall and it is all down to Polly-Tricks.

Once a very fuge long time ago, Mum did a lot and a lot of Polly-Tricking, *and* she was a Cow-Seller. And because she was a different colour Cow-Seller to the Cow-Sellers wot are going to decide about the beach, the Cow-Sellers might just fink Mum's In Charge and hinterfering to cause troubles. Wot she isn't. This time.

Dad is quite actual relieved that Ninja-the-Terry-fried-Terrorist's Mum is going to be In Charge, and so am I. Cos every time someboddedy mentions the beach ban to Mum, she starts ranting about getting-stuff-into-the-Cow-Sellers-fick-heads, and how-can-they-be-so-stoopid. Hopefully, Ninja-the-Terry-fried-Terrorist's Mum has got some betterer ideas and more careful words to use, cos I don't fink bonking the Cow-Sellers over the head with the Pee-Tishon is going to help anyboddedy solve this problem.

None of the Cow-Sellers know that much about Ninja-the-Terry-fried-Terrorist's Mum. She is like a secret weapon. Usually, she is just Auntie Charlotte, a quite very actual plite and helpful hooman, but they has No Hidea how much she luffs Ninja-the-Terry-fried-Terrorist. And how much work and patience she has had to put into helping Ninja-the-Terry-fried-Terrorist be less, well, terry-fried. And terry-frying, to be actual truthful. Nuffink and noboddedy is going to be allowed to hundo that work by taking away hessential and himportant places for doggies to walk.

To be quite actual honest, I are quite very worried for the Cow-Sellers now, cos anyboddedy who can live with Ninja-the-Terry-fried-Terrorist and have a nearly peaceful and happy life must be quite actual Super-Hooman. They isn't going to know wot has actual hit them ...

February 13

Turns out emptying the dishwasher is more himportant than I did realise, cos today Dad went to see the Gee-Pee. He says anyfing is worth keeping Mum happy if it will stop her wittering on and on about it being the 21st century, and Nees shouldn't-just-give-way-with-no-warning and there-must-be-something-they-can-do.

There IS somefing they can do. They can give him a brand noo nee but he doesn't want one: he'd rather put up with the pain and the hoccasional clapsing. Cos the noo nee won't bend a-very-nuff to let him do neeling down. Wot means he won't be able to do sailing or fiddling with his boat. The Gee-

Pee says once Dad decides to stop sailing, he is a Hideal Candidate for a Noo Nee. Mum reckons that Dad will honly decide to stop sailing when he is dead, so he's never going to be a Hideal Candidate; then she stomped off to unload the dishwasher.

february 14

I has decided that it is my himportant work to stop Dad moaning about his Nee, and to give Mum somefing else to fink about other than the Beach Ban and the dishwasher. I has decided that I need to start training Mum and Dad. If they want me to do sit and come-in-their-general-direction and stuff wot they fink is himportant, then they does need to learn some of the fings I want them to do as-very-well.

Last year, Mum tried to get me all very hinterested in a hateful and scaredy fing called a clicker. It wasn't wot you'd call a great actual success. Mainly, I did running away from it, and when that actual Complaint to the Management didn't work, I eated it and left sharp bits hunder the duvet, so that nobobboddedy could fail to hunderstand wot I fort of it.

At first, I fort I could do some woofing as my training actual tool, but I did rule that out quite actual quickerly. Hincesserant woofing is quite actual hignored in this house. I aren't a big woofer or a whiner, but noisy fings like this are HONLY to be used for sig-niffy-cant hemergencies, anyway, like getting stuck in the kitchen with Gipsy-the-Cat. I has trained the hoomans quite actual well that when they hear me doing woofing, it is hessential they do come running straight a-very-way. So I has had to come up with another training tool for fings-I-would-like-to-happen-but-aren't-a-hemergency.

After some fortful finking, I are very quite pleased to hannounce that I has hinvented my own very quite perfick training tool, wot isn't my hemergency bark, and it doesn't go 'click,' so it won't do scaredying anyboddedy. I has called it ... The Biffer!

To be quite actual honest, The Biffer isn't some noo-fangled bit of plastic wot I has hinvented: it's my nose. So far, I has trained the hoomans that when I biffer them, I does want their attention. And biffering them does get their attention very quite well. Hespecially if they is holding a cuppatea in their hand. Dad says I are still a luffly boykin, but his confuser is covered in tea now ... and that isn't actual funny. Even if Mum finks so.

february 15

There is a lot more to the beach than just doggies doing running and playing, Dad says. The people with dogs who use the beach during the winter pick up bits of dangerous stuff that gets washed up there, and also do having a chat with each other. Some senior peoples do walk along the beach and, because they do live alone, being out with their dog and saying hello to everyboddedy is a really, wheely himportant part of their life. And if one of the senior peoples does not turning up, then somebobboddedy goes to bang on their door to find out if they is just having a lazy day or if they has fallen down the stairs.

During the winter time, the honly peoples who go to the beach is peoples with dogs. Apart from one or two very quite HINSANE people who do go swimming. Proper swimming, not my version of playing in the water, wot hinvolves tickling my toes or very hoccasionally cooling my belly in the water, but proper, actual feet-off-the-ground-flapping-your-arms-about swimming. I has got no very actual hidea why they fink this is good, and it does make Mum wince when she sees them. But at least the dog walkers are there to mutter about them being bonkers, cos the Lifeguards honly work during the summer when there is billions of peoples on the beach. But no dogs.

Finking about it, I don't know if I can proper flap-my-paws-about-feet-off-the-ground swim. And I aren't hexactly sure how I can find out, cos the honly way I could do finding out is to try it at the beach. And there is no way I is going to try it in February; hunlike the peoples who do that fing at Southwold, I are not Hinsane.

february 16

I has runned into a fuge problem with my biffer training. The hoomans have all actual got that biffering means 'gimme some hattention,' but after I has dunned biffering, they keep saying "Wotizit-Worzel?" and looking at me like I should suddenly speak Ingerlish or give them writted hinstructions.

Getting off their bums would be a start. Just sittering there and saying "Wotizit-Worzel?" is not what I want to do hachieving, and the honly response I are currently gettering is a luffly stroke and this 'Wotizit' word. Training hoomans is hard work, I do fink.

february 17

You-flipping-Reeky! Today, when I did biffering Mum in the arm every two minutes, she did actual stand up! But then, when I didn't himmediately provide a ten-point plan of action, she satted down again! Honestly, I are a dog wot is doing my bestest here! But I fink this is quite actual progress, although I are sure that training doggies is much, much easier than training hoomans. Perhaps doggies is easier to train cos they don't have confusers or tellys to distract them, and they don't spend every day answering hemails or wondering when they'll get paid.

february 18

Dear Mum and Dad

Between you, you has got eight Oh! levels, two Hay levels, a degree and a CeeEssEee in Art (that's not Mum's, in case you were actual wondering: she's about as good at drawing as I are at singing).You should hunderstand by now that if you gets biffered by me between seven and eight o'clock, it means 'please get my dinner out of the fridge and put it in my bowl.' How hard is that to hunderstand? I has gotted Sit. I has almost gotted Down. I can do Wait and Off-you-go, and even come-in-your-general-direction. Sometimes, I has gotted all these fings without that much troubles. Okay, some fuge troubles, but there is two of actual you and honly one of me. And hoomans is supposed to be the smartest creatures on the planet. Hexcept you two. I has hobviously been dopped by the densest famberly in the world.

From your luffly boykin

Worzel Wooface

february 19
Mum and Dad did a lot and a lot of hugging today, and Mum did crying, and Dad did a bit of hiding his face on Mum's shoulder and putting-on-a-brave-face ... badly.

Tilly has died. Tilly (and Dezzie who camed at the same time) were Mum and Dad's first foster dogs, and the ones that got them quite actual hinterested in Lurchers. Yesterday, Tilly had a special hoperation to find out why she had suddenly started limping, and her forever Mum and Dad, Kate and Giles, founded out she had a horribibble bone cancer that had started to hurt her heart as very well, so they did decide to let her go to the Rainbow Bridge without waking her up from the hoppy-ration. Noboddedy was hexpecting this a-very-tall, and it has binned shocking and upsetting.

When Tilly arrived to be fostered, she had binned used for hunting and fighting, and had never lived in a house before. She was half-starved and poorly sick, and she even had fox teeth that were stucked in her nose. Mum and Dad felled quite very actual completely in love with her.

Hunfortunately, because she had been used for hunting and fighting, she was not cat-safe, and although she was getting betterer and learning how to be in a house, all of the cats went to live in the shed, which meant that, much as Mum and Dad loved Tilly, she could not be dopped by our famberly, wot broked their hearts a little bit.

Finding Tilly her quite very perfick home looked like it might be very actual hard, but, one day, some peoples came to visit Dezzie cos they was hoping to dop her, and, quite to everyboddedy's fuge surprise, Tilly, who could be very shy, marched up to Giles, put her head on his lap, and quite actual clearly tolded the world that she had founded her hooman, and where-had-he-binned-all-her-life?

After some careful finking, cos two dogs was not wot they had planned, Kate and Giles dopped both Dezzie and Tilly, and they went to live together forever in the most perfick home Mum and Dad could have very actual wished for.

I metted Tilly a couple of times, and she was a super bossy booful Lurcher girl. Mum and Dad are so sad tonight, but that must be actual nuffink to how Kate and Giles and Dezzie is feeling. Tilly had five perfick years in her forever home, and everyboddedy here is trying to actual concentrate on that fort.

february 20
The previously ginger one is Hofficially my star actual pupil. And I didn't heven need to biff her. She sawed me biff Dad, and then she said "Wotizit-Worzel?" and I forted fings were going to get heven more complercated with me having to deal with three stoopid hoomans. But instead she did saving the hooman race by saying "Does-he-want-his-din-dins?" in a quite very actual silly, soppy voice wot works with the cats. But I did exerlent hignoring of this and followed her into the kitchen, and then she did Give. Me. My. Dinner. And even betterer, she did tell Mum and Dad she had dunned this fing, so I are hoping that they will have gotted the very message at last!

february 22

So, yesterday, I had to biff Mum again and again, and a-very-gain, but I has succeeded. I was a quite very exerlent boykin: I did not give up, and I did not get too very actual frustrated. I might have biffed a bit harderer towards the end, though, cos I was gettering quite actual hungry. After about 17 billion biffs, Mum gave up trying to write her hemail, stooded up, and gotted my dinner out of the fridge. Mum now hunderstands Biff.

And when I started biffing Dad tonight, Mum did tell him that Worzel Wooface wanted his dinner, and wasn't going to stop biffing him until he stooded up and got the chicken wings out. So he might as well do it now before he ended up with his cuppatea in his lap.

february 23

Biffing is not just for dinner, but it is a very quite good start. Currently, the hoomans fink it is honly for dinner, but I is working on a plan to get them to hunderstand it can mean different actual fings. I has just got to work out some more hinstructions wot they will hunderstand.

I may be some time ...

THE BIFFER

Most people fink my nose is made
For smelling and for sniffs
But it has got another use
It's great for doing biffs!

I biff my Dad when I am bored
Or want to let him know
I'm starving blinking hungry
And it's time for him to go

And get my dinner out the fridge
Quite actual quickerly
Or I'll biff him in the arm again
And spill his cuppatea

february 25

Mouse has losted a tooth. One of the sharp, pointy scary ones at the front. Mum's feeling quite actual bad about this cos, for the last couple of weeks, Mouse has binned dribbling when she has comed for a cuddle. Mum putted it down to Mouse being weird and over-hexcited about being stroked, but yesterday Mouse did a fuge yawn, and we all saw there was a tooth missing. Now Mum's fort about fings, she says she hasn't very actual looked at the hinsides of any of the cats' mouths for, well, ever.

So, today every cat is going to get hambushed and hinspected.

I wish Mum had decided to do this fing before Dad wented to work. Being down the harbour seems like a quite actual good idea right now. I don't fink there's anywhere hidded a-very-nuff to escape wot's going to be happening here ...

february 26
****FUGE NEWS****

At the end of last year, we did get some noo actual neighbours. At first I did fink that Kes, Maisie and Kite (who is all Labradors) did live in the house all the time, but turns out that Kite and her Mum and Dad were just doing a lot and a lot of visiting. But now Kite and her Mum and Dad have moved in forever.

I are quite actual pleased about this, and so is Mum. Ever since Pip and Merlin did moving to the next village, Mum hasn't had a regular walking person to do chatting with whilst I do running about. And noboddedy to distract me from bogging off. Kite's Mum is quite very keen to do walking together and, most himportantly, Kite has got a perfick recall and never, hever does bogging off. So now Kite's Mum has moved here forever, we is going to get horganised and do regular walking stuff together.

I fink this is fabumazing. Mum has not tooked me for a walk anywhere but the beach since my haccident, cos she's nervous and worried about me bumping into stuff, but she finks with Kite being a good hinfluence, and Kite's mum stopping her being redickerless, it might be quite actual okay.

february 27

Today, me and Kite, Kes and Maisie went for our first walk together since their mum did moving here forever. I fink we will be allowed out again together cos I did not do bogging off or falling into any ditches. This was mainly because we did find some fuge puddles wot me and Kite roared up and down in, splashing everyboddedy, hespecially Maisie, who was doing her bestest wallowing-like-a-hippy-potty-miss. Dad says saying Maisie is like a hippy-potty-miss is a bit actual rude when we is honly just making friends with the noo neighbours, but I do promise I wasn't comparing wot she does look like, just wot she actual does, which is wade into the middle of a puddle and lummox about. I did have to do some quite very hathletic jumping over Maisie, but it was that or land on her head, and it did hinvolve a lot and a lot of splashing of Mum.

Splashing mud isn't wot anyboddedy does fink is a reason for not going on walks in my famberly, just bogging off and chasing fings and falling into ditches, and I did none of those fings, so I has binned promised another walk tomorrow.

How-very-ever, after we finished our walk, I did hear Kite's mum talking about a Quick. Hose. Down. Wot, as far as I are concerned, is the Worst. Fing. In. The. World, and I nearly decided to go right off Kite's Mum.

Fortunately, Mum did very quick hexplaining, with a lot of sighing, about Worzel being scared of the spitty yellow snake fing, and that a Quick. Hose. Down would result in a Fuge. Melt. Down. She tooked me home to deal with the Mud in my own way, wot mainly hinvolves lying down in my bed and waiting for it all to fall off. Or jumping on the sofa when noboddedy is lookering.

March

March 1

The previously ginger one wants to Change the World. I do find this very quite surprising, cos she has binned hiding in her bedroom for most of the last week, and hasn't been able to change herself from lying down to standing up. Maybe she ought to try that fing before moving onto more quite very actual hambitious changes.

She didn't do standing up for very long today, but did manage sittering up, and ever since she has binned tapping away on her Confuser, and also telling me stay out of her room. I are not allowed to wake up Mouse cos the previously ginger one says she can't do her himportant work *and* deal with Mouse yowling at her or trying to stand on her head. On the previously ginger one's head, that is. I don't fink Mouse can actual do headstands, but then I didn't fink the previously ginger one was going to sit up today, so I could be actual wrong about that fing ...

March 3

Today, I did have a noo hexperience! It is a quite very actual fact that I is a luffly-ever-so-Umble-boykin wot is respectful to other doggies. And I do have good manners, wot is hessential, cos I do meeting lots of other dogs.

In my first year in my forever home, Mum did make sure I honly had positive hexperiences with other dogs, because I was actual anxious and easily hupset. After a quite very long time, I was allowed to meet Ghost, who could be a bit bossy and barky. You might do remembering that Ghost did die last year, wot was a fuge shock for everyboddedy. Well, today, I did meet Marco, who is Ghost's brother-wot-he-never-metted. He is a Gee-Ess-Dee, just like Ghost. He is honly seven months old, but already as big as me! He is going to be fuge! Today, it was my himportant work to be the good hinfluence and positive hexperience for Marco. Wot means I are hofficially growed up and confident to do that fing!

I didded my very, very best, and there was lots of careful watching by Mum and Claire, Marco's Mum. When it was just me and Marco, everyfing was quite very actual good fun. Marco did lots of playful stuff, and I did quite very gentle, not being reactive. Or too quite actual bossy. And we did lots of running round and round in actual circles!

I did find it much harderer when Nelson (who is Marco's ickle big brother) was allowed to come and join in. I does not like having one dog sniffing my gentleman bits and a-very-nother trying to sniff my mouth, so I did get a bit flighty, and belted off to have my own actual space.

But because Nelson is a good boy at doing 'come,' and other stuff like that, his Mum did calling him to her so he couldn't do his 'sertive 'that's my brother you is saying hello to, you be a good boy or actual else.'

I are quite actual tired now, and there was lots and lots of learning for me and for Marco, but it was a big success and Mum says I are a super, luffly boykin.

March 4

I aren't that keen on March. One day it seems like it's going to be a luffly day with a bitta warm sun, and then the next day, it gets quite actual icy and chilly, and I-aren't-going-out-to-do-a-wee-in-THAT-fanking-oo-kindly.

I does prefer January and February to be quite very honest. You know where you actual is with January and February – hunder the duvet in the warm, and not getting tricked by some sudden sunshine wot makes you all oppymistic. March is quite very actual confuddling and tricky, I do fink.

I are not the honly one confuddled by March: our house heating can't hunderstand it, neither. Our house heating keeps trying to guess what the weather will be like but keeps getting it wrong. With March boinging about all over the place, and sometimes being warm and sometimes – like today – flippin' freezing, the house is either very actual chilly or feels like somebodddedy's set fire to the stairs. And then the next day, it realises it did over-do the roasting it gave everybodddedy, and turns itself down ... by which time, the wind has started blowing in the other actual direction, and it feels like somebodddedy has lefted open the back door.

Wot they haven't cos Dad's binned and checked about a billion times, and shouted up the stairs about who's-got-a-window-open? And stomped around and closed all the curtains, and moaned, and put on so many jumpers and his dressing gown, and he is still not warm. So now Dad is sulking, and going on and on about whether the house heating fermostat is broked (wot it isn't, Mum says: it just can't cope with the up and down and hot and cold weather at the moment; it does this every year). There's actual nuffink she can do, and we've all just got to put up with it.

THE LIE-IN

I don't like the weather, it's cold and it's wet
The wind is too ruffly and it seems to get
Right down my hearholes, right up my bum
So I'm staying in bed, and so is Mum

I don't like the ice, and I don't like the snow
when it blows all about in my face, although
The sun's coming out now, wake up, snoozy head
I want to go out please, get out of your bed!

It's time for a walk now, I know it's still chilly
Under the duvet, you can't hide - that's silly

It's a booful crisp day, you can stand in the sun
Get dressed Right Now, it's time for my run
You're being quite stubborn, I fink I might need
To woof in your face if you don't put on some speed
You've got gloves and wellies, and here is your hat
I'll even do Sit for my lead, on the mat

Sorry for waking you up this fine morning
What do you mean, you'd prefer to be snoring?
You knew the rules when you chose to have me
You can't have a lie-in unless I actual agree

March 6

Dad says he's Dot a Dold. He hasn't got Flew cos Flew is a serious illness wot Dad hasn't got, however much he finks he is going to die, Mum says. He's blaming the house heating for making him ill. He finks all the sudden heating changes have created the Perfick Breeding Ground for Bugs. He keeps croaking that one minute he feels cold and the next minute he's hot. And can he have a cuppatea?

I aren't sure where the cuppatea fits in, but I fink Dad's ferm-O-stat is broked as well. Mum says there's nuffink she can do about Dad's dold, neither, and we're all just going to have to put up with that, as well.

March 7

The previously ginger one has worked out how she wants to Change the World, and as a very actual bonus, it will help her to pass one of her college hexams, wot Mum reckons is probababbly more himportant, right at this very moment. Mum is saving most of this talk for Dad, and being very fusey-tastic about the previously ginger one's Fight The Stigma Project, wot is the fing that is going to CHANGE THE WORLD. I has got no hidea wot a stigma is, but if the previously ginger one reckons it is himportant a-very-nuff to get out of bed and go to College and have a fight about, then it's probababbly somefing Orrendous that does need fighting about. Cos very nuffink has binned that himportant to the previously ginger one, apart from sleeping and cuppateas, for a long, long time.

March 8

I've runned into a fuge problem. Whilst I has binned learning about my Noo Neighbour Kite, Mum and Kite's mum has binned gettering to know each other as-very-well. And they has got Somefing In Common.

Fings were going along quite actual nicely when me and Kite were the honly himportant fings they knew they had in common, but now we has got compertishon ... and I are not himpressed a-very-tall.

Turns out, Kite's mum likes gardening as much as my Mum likes gardening, wot is a disaster as far as I are concerned. The last fing I need is someboddedy hencouraging Mum to be even more hinterested in gardening, and less pleased with me diggering up her plants wot are in-my-way-of-finding-the-mole. Now the sun is coming out very sometimes, all they can talk about is wot they will be growing, and the changes they has got planned and ... it's hendless. On and on and on, they keep yackering about it.

March 10

Me and Kite has decided it is our himportant work to stop all this gardening talk, but we has got different actual hideas about how to hachieve this fing. Kite does exerlent staying off the flower beds and pretendering they don't hexist. My feery (wot I do fink is betterer) is to jump all over Kite's mum's flowerbeds so that everyboddedy does decide that I am squashing them, and that going out for a walk is a betterer hidea than playing and roaring round Kite's fuge garden.

It's not working. My feets aren't quite big a-very-nuff, and I don't have Maisie's fuge belly or Kes' clumsiness, and I aren't causing enuff chaos to make any actual difference. All that happens when I stand in the flowerbed, is that I get asked to 'geroff,' and then Mum and Kite's mum go over to see if I've dunned any damage. Then they spot something hinteresting that has started to grow, and start yackering about it, and does-Mum-want-a-bit?

Wot she does quite very actual NOT. Me and Dad do both actual hagree with this fing. We has got far too many blinking plants in our garden. And there is a fuge heap of plants waiting in pots already without Kite's mum offering even actual more. There is No. More. Room. in the garden as it is. Dad says Mum's going to ask for more flowerbeds soon, wot is hateful work. And I still haven't found that blinking mole ...

March 11

Kite's mum has gotta Greenhouse. We don't have a greenhouse. Dad says he's going right off Kite's Mum ...

March 12

Somefing fabumazing happened quite very recently. I gotted to do catching up with my foster sister, Hazel-Hattie Princess Poo Pants.

When Hazel did come to stay here a year ago, she was a very actual poorly girlkin wot did have mange, and a tail wot looked like she'd borrowed it from a actual rat. We called her Hazel (and Princess Poo Pants cos, well, she was) but now she is called Hattie. Hattie-Hazel got betterer and fugerer, and now she is a super fast girly, and can actual run fasterer than me on sprint, wot is quite actual shocking! I has decided that it is very okay, though, cos she's my foster sister, and I can do being proud of her instead of feeling put out.

We gotted invited by Hattie-Hazel's famberly to come for a play date at Norfolk Dog Facilities, wot was very safe and hinteresting, with lots of rabbit smells and mud. They has builded a very special mud comer, just for Worzel Wooface, cos they knewed I was coming and would actual need this himportant facilerty. Hattie-Hazel and I did have a fabumazing time but, being foster-siblings, we did take the whole bitey-facey game fing as very far as you can. Honly siblings really, wheely know how to do proper play fights, don't they? Mum's never seen me doing jumpering up like I did today, so it was all a fuge treat for her as very actual well.

March 13

Dad says we Do. Not. Have. Room. for a greenhouse. Not even if we chop down the Napple Tree. Mum's sulking now cos she wants to grow a naubergine, wot is a funny purple vegetababble.

Dad doesn't like vegetababbles, hespecially vegetababbles wot hinvolve him digging up half the garden and chopping down trees.

March 15

The previously ginger one's Fight the Stigma project is all about Men-Tall-Elf. I fink it is probabably about very time I did some hexplaining about the previously ginger one, because if she is going to Change the World and Fight the Stigma of Men-Tall-Elf, then I better had do my bit as well.

And also she is 18 now, wot makes her a hadult, and she says I can do that fing. The previously ginger one might have troubles with her Men-Tall-Elf, but she is just as quite very bossy as Mum sometimes. Mainly that bit, if I are quite actual honest ...

Hanyway, the previously ginger does struggle with a Men-Tall-Elf wot causes all kinds of problems for her, like feeling worried and stressed and tents, and also very low. Not sad, hexactly, but like she is being squashed by a fuge, foggy cloud. And sometimes, the cloud does get so actual heavy and difficult that she can't do gettering out of bed. Or even want to go on living, wot is hawful and difficult for everyboddedy, but most hespecially her.

Living with a Men-Tall-Elf inside your head must be very hard blinkin' work, I do fink, and she does take some med-i-sins that help, but her Men-Tall-Elf isn't the honly fing that is himportant about the previously ginger one, who is bossy and quite actual clever, and probababbly the bravest person I do know.

She must be cos she's decided to tell as many peoples as she can about wot it is like living with a Men-Tall-Elf, and actual hencourage everyboddedy else who has a Men-Tall-Elf to do the same fing. And then peoples who do not have a Men-Tall-Elf will realise that they is mainly just normal hoomans like everyboddedy else, and that will help Fight the Stigma.

Mum and Dad have dunned a lot of talking with the previously ginger one about is-she-sure-about-telling-peoples? and once-you-say-somefing-you-can't-unsay-it, wot sounds blinking hobvious to me, but somefing hoomans need to do finking about a lot and a lot. But the previously ginger one says she's certain and sure, and Mum says that is brave and brilliant, and she'll help if she is needed. Dad says the previously ginger one is just like her somefing-beginning-with-B mother, but he had a fuge smile on his face when he said it, and his hiballs kepted blinking a lot and a lot, so I fink he is quite actual proud.

March 16

Kite's Mum says my Mum can grow a naubergine in *her* greenhouse wot is very quite kind of her, I do fink. And also means Dad can stop hiding down at the harbour, pretendering to fix fings on the boat ...

Dad says I shouldn't go back to snoozing and we should be on Higher Lert. The greenhouse fing might have gonned away, but it Isn't. Over. Yet. He's binned living with Mum for a long actual time, and he finks she's on one of her Progress Missions, wot is going to be hexpensive or hexhausting or actual hannoying for at least one of us.

March 17

I think Dad was right. Mum says that if she can't have a greenhouse she's going to do Somefing About Dad's Shed, to see if she can make some room in there for her to grow her plants and seeds. Dad did fink about doing Complaints to the Management, but he didded some complercated sums in his head instead, wot went somefing like this –

$$\frac{\text{Chopping down the napple tree} + \text{building a greenhouse} + \text{time} + \text{moneys}}{\text{Mum fiddling with his stuff} \times \text{finding fings he promised to chuck away}} = \text{'Yes dear' with a lot of sighing and looking hexasperated}$$

March 18

****PROGRESS****

‡ Progress is when fings take a step in the right direction, but have got very nuffink to do with walking, and is mostly about decky-rating or gardening

‡ The honly person who is really that hinterested in progress is Mum

‡ Me and Dad are mainly hinterested in an Easy Life. Progress and an Easy Life don't go together

‡ It is himpossible to have an Easy Life when Mum is in one of her Progress Mission moods

‡ Whilst Progress is happening, it is best to do joining in. If you don't want to do joining in, it is best to hide

‡ Dad is betterer at hiding than me

‡ Sometimes, Mum decides that I must make progress. On fings like come and sit

‡ Then I am betterer at hiding than Dad

‡ There is not a-very-nuff room for me *and* Dad to hide under the bed. We tried

‡ And anyway, he has a boat

March 19

Dear Small White Teddy Bear wot looked like a hescapee from a four-year-old's actual bedroom.
Fank oo for making Sunday lunch in the pub, when the hoomans were doing talking for Ingerland, quite very actual more hinteresting than I forted it was going to be. I did fink I was going to do lying on my blanket and getting bits of roast beef sneaked to me by Dad. Which is quite a good actual way to spend a Sunday lunchtime. But then you did turn up, and wot was going to be a snoozy-nibbly sort of lunch turned into a edercation.

You was so fusey-tastic about saying hello to me, I forted you were pretendering to be a hooman. You did waving your front legs around all hexcited like Mum does sometimes, and also lots of squeaking.

When I do meet ickle dogs - wot Dad did finally convince me you actual are - I very usually do helpful and considerate lying down, so we is all on the same level for hellos and plite sniffing. You did not want to do that fing. You wanted to stand up on your back legs and show me your bestest being-a-hooman-dancing. The whole time. I don't fink you putted your front feets on the ground the hentire time you was in the pub! I has never seen somefing so hexhausting in all my life.

I don't fink Mum and Dad would like it if I didded dancing on two legs cos then I'd be taller than them. Wot would be strange and the bits of beef wouldn't be putted in the right place a-very-tall. They'd have to get stucked on the ceiling. And not putted on the floor. Dad did decide that you was very actual sweet, and wented to have a cuddle and a sit-on-the-floor-hello with you. Mum finks Dad is a big softie, and maybe he needs more cuddly toys on his bed at night, but that isn't actual very manly so I probababbly shouldn't do saying that fing.

You was so very friendly and fusey-tastic that I did not feel worried or concerned about your very hexcited greetings, wot Mum did have a small moment of panic about, hespecially when she did remembering that I have a very ex-teddy bear in my toy collection wot looks exactly like you, hexcept I has mostly pulled off its head... But I did not do that fing to you. And Mum says I was a luffly, friendly but very quite bamboozled boykin.
From your luffly boykin
Worzel Wooface

March 20

The cats has binned really, wheely scaring me today, hespecially Mabel. Just recently, Mabel has decided she's going to be a Normal Cat: one of those cats that does coming hinside, and maybe sometimes sitting on a hooman and not hiding in the shed waiting for the world to actual end.

This is definitely good news for Mabel but not for anyboddedy else; hespecially me. Mabel's idea of being Normal mainly hinvolves her hissing and spitting at all the other cats, so she can be Normal without them watching her. Mabel wants to be Normal, but she wants space to do it in. All the space in all the house, to be quite actual honest, so all the other cats are getting squashed into smaller and smaller spaces, out of the way of Mabel.

Mum finks Mabel is being Normal cos of all the work we did with Mabel to stop her being terry-fried of everyfing. Dad reckons it's cos Mum is crashing about in his shed doing tidying up and moving stuff around, and Mabel's safe space has becomed very unsafe. Really, wheely unsafe, with fings falling off shelves and getting dragged about, so Mabel has got no choice about being Normal currently. It's either that or become homeless. Or shedless, finking about it.

March 22

Mabel has robber-dobbed Gandhi's carefully-hagreed-with-all-the-other-cats day sleeping place. So now he's nicked Gipsy's day bed, which she is quite actual cross about, and On. The. Prowl. for a new sleeping place. Mum decided that Gipsy – cos she is the senior cat round here, and did not react too very badly to Gandhi and end up costing her moneys at the vet – did deserve a special noo bed. It's wicker and wooden, and probababbly cost someboddedy a flipping fortune – but not Mum: she got it for nuffink.

She found it at the dump when she was making Progress by chucking away lots of cardboard boxes wot was hempty, and a couple of cardboard boxes that weren't very actual hempty, but full of redickerless junk that Dad. Doesn't. Need. And wot Mum is quite actual certain he has forgotted about. And we-won't-be-telling-him-will-we-Worzel?

After Mum got back from making Progress at the dump, she did give the noo bed a proper good scrub, and founded a booful cushion for it, and then, with a quite actual proud and pleased 'ta-da!' she putted it down on the shelf where Gipsy has binned sulking since Gandhi stoled her sleeping place.

Gipsy hates it. Not a-very-nuff to let any of the other cats sleep in it, but she has decided to lie right beside it, making it perfickly clear she does not find it hacceptababble, and giving it disgusted looks like it's somefing that got dragged out of, well, a dump.

I fink Mum is a bit actual hoffended now; she's gone out to the shed to make more Progress and I fink she might take it out on Dad's junk.

March 23

Gipsy is still actual refusing to use her posh noo bed, but isn't lettering any of the other cats use it neither. Today, Mum came in from the shed carrying a muddy cardboard box, and saw Gipsy lying on the hard shelf, wot can't be good for her old bones, so she dumped the muddy box on the shelf whilst she nipped upstairs to get a soft blankie from the fuge ginger boyman's bed. To see if Gypsy preferred that.

When Mum gotted back from collecting the soft blankie, Gipsy was nowhere to be actual seen! After calling Gipsy all the somefing-beginning-with-B words she could actual think of, Mum started to drag the fuge muddy box off the shelf, honly to hear the box make some 'leave-my-noo-bed-alone' ass-matic squawks.

This is Not. Progress. Mum says, hespecially if Dad sees wot's in the box.

March 24

I fink I has had a-very-nuff of Mabel being Normal.

My route from the big bed to the kitchen used to be –
'Hello Gandhi' at the top of the stairs
Run like mad past Mouse in case she's having one of her moments
Very carefully and respectfully hignore Gipsy in the hall
See wot Frank has chucked off the table onto the kitchen floor
And most very actual Don't. Look. at the kettle where Mabel usually sits when she is waiting for her somefing to eat

Since Mabel has decided to be Normal, though, none of the cats is anywhere I can do predictering, and my saunters round the house have turned into a bad hepisode of *Run The Gauntlet*, cos the cats keep appearing in places I aren't hexpecting, like behind the sofa or inside a laundry basket, where they has got no actual business to be hiding. And definitely NO business attacking me for disturbing them when it's Mabel wot has caused all the troubles in the first place.

March 26

Mum's still trying to make Progress with the shed, and the muddy box that Gipsy has decided is her noo bed isn't hexactly the helegant and stylish accessory she wanted on the shelf.

And worstest of-very-all, when Gipsy wasn't in it the other day, Dad had a look inside and gotted all hexcited, like it was his birfday. He has spended the last few nights re-actual-discovering all his precious fings that Mum was planning to take to the dump. It is not precious, Mum says, it is junk. Dad says it isn't junk, it's His. Stuff, and Mum has no hidea how himportant cassette tapes with the shipping forecast from 1983 might be in the future. Or the lead from the thingyamewotsit that he can't find. And he won't find it, neither, cos that went to the dump a week ago ...

Dad's started hoarding all his precious stuff in front of Uff-the-Confuser. Mum says this is even actual worse than Not Progress: this is a Disaster.

March 27

Last night, my bad hepisode of *Run the Gauntlet* tooked a noo direction. First of-very-all, I got bopped by Gipsy, and then Mouse ranned down the stairs in hystericals. When Frank and Gandhi hadded a wrestling match on the hall floor, wot I was not allowed to join in with, even though it looked fabumazing, I did find my-actual-self so very over-hexcited and worried and stressed and wobbly, that I forgotted the Number One Rule of Mabel ... and chased her out of the cat flap. Mum was not pleased. A-very-tall. Mum might not be that actual pleased, but the other cats is very quite relieved.

March 29

Dear Mum

You know that bitta fish you has binned trying to get me to eat, wot I have dunned careful hignoring of every time you do put it in my actual bowl? You will be pleased to actual know I has dunned eatering it. Can we go back to chicken wings now?

From your luffly boykin

Worzel Wooface

Dear Mum

You know that bitta fish you has binned trying to get me to eat, wot I did do careful hignoring of every time you put it in my actual bowl, but then did do eatering it? You won't be so pleased to know I has dunned frowing it up all over the bed. Can we go back to chicken wings now?

From your slightly less dorable luffly boykin

Worzel Wooface

Dear Dad

You know that bitta fish Mum has binned trying to get me to eat, wot I did do careful hignoring of every time she put it in my actual bowl, and you has binned telling her to give up about? Well, I did do eatering it, but now I has frowed it up all over your side of the bed. Sorry about that fing. Can you tell Mum it's time we wented back to chicken wings?

From your currently-in-the-doghouse-luffly-boykin

Worzel Wooface

March 30

The clocks has gone forward. Time and clocks is too complercated as far as I is concerned, and None of my Business, but the clocks going actual forward means that Dad is quite very happy. Dad don't really care if it is hot or cold or raining or hailing, or that the clocks going forward mean that we might soon see some warmer days, he just wants the *longerer* days.

It's nearly sailing season, wot means he's got to fix all the bits of the boat he hasn't been able to do in the dark when he's finished work. And also fix all the bits he broked in the dark when he couldn't see properly, but couldn't resist fiddling with to see if he could do it. Wot he couldn't. Wot got hexpensive. Wot he hasn't tolded Mum about.

March 31

Fank flipping goodness for actual that! Mum has finished the Progress in the shed. She is quite very pleased with the Progress, and now that she has stopped crashering about in there, Mabel has decided that being a Normal Cat is quite over-actual-rated, and has decided that being a Mad Cat in the shed is actual much easier. Everyfing is very actual posh and horganised in there now; even halmost stylish, hespecially now that Mabel has her very own quite booful posh noo bed.

Fings aren't quite so posh and stylish in the house, though. Gipsy won't let Mum swap the fuge muddy box for a cleaner one, and the space round Uff-the-Confuser looks like the last half-hour of a jumble sale, when the honly stuff left to buy is the fings noboddedy wants, or knows what to do with.

Mum says she's going to chuck away one bit of Dad's jumble sale junk every day; he'll never notice, and we-won't-be-telling-him-will-we-Worzel? Again. But she has very wisely decided she is going to have to live with Gipsy's muddy box on the shelf, wot I do fink is very quite sensibibble. She hasn't got a hope of getting that cardboard box away from Gipsy: she's far too blinkin' clever ...

Visit Hubble and Hattie on the web: www.hubbleandhattie.com
www.hubbleandhattie.blogspot.co.uk • Details of all books • Special offers
• Newsletter • New book news

APRIL

April 1

I are trying quite actual hard not to cast nasturtiums on my famberly, but seeing as today is April Fool's Day, I do fink I should tell you how my Biffer training has binned going.

Fick and Foolish are halmost the same actual fing, I do fink, wot does probababbly give you a hidea of how fings is getting along. I fort back in February that the hoomans woulda made more Progress than they actual has, but wot with Mum doing her own Progress in the shed, and Dad doing his bestest to actual havoid being dragged into Mum's Progress, my own actual himportant Progress-with-the-Biffer has binned hignored. I has tried to put this down to them being tired and busy, but I can't seem to get them any further than Biff equals Dinner. I has hexperimented biffing with different patterns but they aren't gettering it. A-very-tall. I might just have to do giving up and accept that they is Fick. And fools.

April 3

When the previously ginger one did say she was going to tell everyboddedy about wot it is like to live with a Men-Tall-Elf, she was not actual kidding. Somehow she has managed to get hold of the Head Hitter of our local noospaper, and he has dunned telling the hole world (well, everyboddedy who lives in our area) about her campaign, and about her Men-Tall-Elf.

To be quite actual honest, the previously ginger one wasn't all that hinterested to know wot the Head Hitter had written. Happarently, the noospaper did show it to her before they printed it. How-very-ever, she had not seen the photos wot they had tooked, and she did spend a long-very-time asking Dad if she did look fat in the photos, and whether her lipstick did match her skirt, wot he isn't hexactly a hexpert at discussing.

Dad reckons she might want to Fight The Stigma and Change the World, but she wants to make sure she looks perfick whilst she's doing it. Mum says she wishes she had known before the photos got tooked that she was very hexpected to be in one of them, as she woulda dunned a lot more actual concentrating when she got dressed that morning.

April 4

We has Friends in the North. I did not know this fing, but me, Mum, and the previously ginger one have been hinvited to go and visit them, and we will be leaving tomorrow. I are pretty actual certain there is a quite very fuge lot of North in the world, so I are refusing to get fusey-tastic about visiting Our Friends in the North until more hinformation is gived to me. Mainly, I want the answers to two very himportant Questions –

* How far North?
* How are we gettering there?

With my famberly, it could be by boat, train, car, or even on foot! (The last one would be quote actual favourababble, although it does depend on the answer to the first question.)

April 5 (early)
Answers to my Questions –
*   Blinking flipping miles and miles and miles
*   In the car
It feels like we has binned travelling for months, possibibbly even a hole year. So far we has stopped off for coffees and weesandpoos twice, but I did decline to have a snack. It is very quite himpossibibble for me to do hexpressing all my Complaints to the Management about this very long and quite hawful boring journey, so I has dunned one fuge Complaint to the Management, and I are on Unger Strike. I are pleased to say that this has binned noted by Mum and the previously ginger one, and there has been lots of talk about poor-Worzel-it'll-be-worth-it, and also trust-me-you'll-love-it-when-we-get-there.

Currently, I are sulking and planning to Not. Trust. Mum. ever, hever again. Or at least until we finally reach our Friends in the North, whichever comes first. Wot I do hope is quite very soon, cos I are getting a bit peckish now, and I am beginning to regret my actual Unger Strike, cos I are feeling quite actual car-sick.

April 5 (several years actual later, I do very fink ...)
As part of the actual deal for me gettering in the car and staying in there for-HEVER, Mum said we would be stopping off at Greyhound Gap for weesandpoos, and so I could do stretching my legs so I was not too fullabeans when we did arrived to stay with our Friends in the North.

Greyhound Gap is another charity wot helps Greyhounds and Lurchers find forever homes, but unlike Hounds First, Greyhound Gap has got a kennels where really hupset and poorly dogs do get a chance to get betterer, or at least learn some manners before they go to foster homes, or their forever home. They do also have some dogs wot will stay at Greyhound Gap forever, cos they has got Special Knees, and wouldn't be happy in a forever home.

Whilst we were at Greyhound Gap, I did have a chance to say hello to Lisa, who is very actual In. Charge. of Greyhound Gap, and another one of them Hexperts, just like Gran-the-Dog-Hexpert. I could tell this fing before I even gotted out of the car cos she was wearing wellies and a muddy t-shirt, and looked like she did not fink brushing her hair was a pri-orry-tree. Mum says Lisa is the single most terrifying person she has ever, hever met. I did not hunderstand this a-very-tall: I did fink Lisa was one of the most easiest-to-hunderstand-people, and quite actual luffly. She was very horganised, and didn't flap her hands about, or try to touch me at all. I wasn't really hinterested in the words coming out of Lisa's mouth, just the fings that she did. Mum says it was a quite actual good job I wasn't hinterested in the words coming out of her mouth, cos some of them weren't suity-ball for plite comp-knee.

At Greyhound Gap there is a field where luffly boykins can do safely running about, and after I had settled into the field, Lisa did say that Rio, who is a clever and shiny black Lurcher girl, could do joining me. And show Lisa that her special secure field was mainly, sometimes, usually secure, unless you is a Rio and can jump higher than you can ever himagine. How-very-ever, the hole of Greyhound Gap has a even higher fence that even Rio can't get out of, so Lisa did a lot of putting up with Rio's jumpering, and said a lot of words-beginning-with-B about her really, wheely needing a hagility home. When I did hear the words hagility, I did skulking off to have a concentrated and himportant time sniffing in the hedges, where I fort it was safe that Rio wouldn't land on top of me when she camed flying over the fence.

Fortunately, both Lisa and Mum did very hagree that I was not a hagility dog, and there was a lot of sighing and muttering about it taking two years to persuade Worzel Wooface to try anyfing noo. I are quite actual pleased about the Worzel-not-keen-on-hagility words, and mainly trying to remember this bit of their talking, not the bit about me being too anxious and Not. That. Clever. I are perfickly clever, fanking oo kindly; I just don't do clever fings hoomans actual hadmire. I do clever fings wot Worzel Wooface does actual hadmire, like havoiding hagility. And other redickerless fings like that. Getting back into the car was hard a-very-nuff, fanking oo kindly, but that had actual nuffink to do with jumpering or being flexibibble; more to do with the fact that I'd had more than a-very-nuff of car journeys.

April 5 (Finally ...)

Sometimes, hoomans can be actual quite stoopid, and miss out Himportant Bits of Hinformation wot might actual stop a luffly boykin losing the blinking will to actual live. And I aren't talking about going in the car and driving so far North you do start to wonder if you has started to go South. That bit, I do fink, was probababbly wise not tellering me about, cos I would not have gotted in the car in the first place. But ... very actual BUT ...

Noboddedy mentioned Kelpie and Finn. I does not know how in Heckington Stanley this was forgotted about. Or considered Not. Worth. Talking. About. So now I shall do correcting this Orrendous mistake.

Kelpie is a booful and helegant very, very senior hairy lurcher. She is sixteen years old but can still do running about.

At-very-first, she was a bit actual wary of me, but after I did exerlent hello-hincredibibbly-himportant-Queen-Kelpie, and I-are-a-humble-boykin-nuffink-to-be-halarmed-about, she did decide that I was very welcome in her house. I aren't completely surprised that Kelpie did do some fortful wondering wot I was like, and if I was plite and gentle, cos she lives with Finn, who, to be very actual truthful, is as mad as a box of frogs. With cherries on the top.

Finn is a Nirish Lurcher, wot founded himself with Our Friends in the North after a complercated journey across the sea. It is quite very clear that Finn has decided to actual hembrace his Nirish roots, cos he is a hexpert at Nirish Jigs. He is a Whippety-sized Lurcher with tiny ickle feet wot click and clack, and he never, hever stops moving! Even when he is lying down he does quiver with hexcitement. I do fink he should do a naudition for that Raver-Dance show: he would be actual fabumazing!

There hasn't been a fuge amount of time for playing this evening after our redickerlessly long journey, but I has ended my Unger Strike, and just about actual forgiven Mum for the car journey. I do fink I might start trusting her again tomorrow. Maybe it *was* actual worth it, after-very-all.

April 6
Our Friends in the North has got a Rabbit Fee-at-Her. It's fabumazing, and very actual hinteractive. And in the garden. Luffly boykins wot sit quietly and gently and very hunder supervision are allowed to watch it for hours, if they do want to. When you is all settled and comfy, the show begins, and two rabbits come down out of their little house and walk down a plank fing. It is very quite hessential that luffly boykins do keep hold of their hexcitement when this happens, cos the the rabbits do come onto the grass and eat their breakfast. And if you is completely still, they do come to the edge of the fence ... and Give. You. A. Kiss. I aren't kidding! I has no hidea how I did manage to keep perfickly still and calm, but I did. I was perfick.

Finn wasn't. Whilst I was watching the Rabbit Fee-at-Her, Finn did decide that my boddedy was a cushion, and tried to do bonkering me. It is quite very almost himpossibibble to do Being Quiet and Not. Make. Sudden. Movements when another doggy is pretending you is a cushion for bonkering, but I did do my bestest, and Mum and Our Friends from the North did quite a very lot of hembarrassed tellering Finn to stoppit, and come-away, and oh-look-here's-a-ball. Finn wasn't that actual hinterested in the Rabbit-Fee-at-Her cos he'd seen it lots of actual times before. I was so actual quite hinterested in the show, I did not do a-very-nuff telling him to bog off with knobs on, and now he finks he has gotted permission to bonk-me-like-a-cushion. All the time.

April 7
Tomorrow, Mum says we is going on a Fer-eee cross the Mer-zeee, wot if you is a hadult who is old a-very-nuff to remember the song, you can't actual say, you has to sing. All evening, Mum's binned talking reasonababbly about fings, but then getting to the Fer-eee cross the Mer-zeee bit, and singing it loudly and badly. The previously ginger one says that if Mum doesn't stop her racket, she'll be going on her own, cos she's got a horrid feeling Mum is going to do the Fer-eee cross the Mer-zeee howling when we is actual on the boat. She says she's very glad she isn't a hadult wot was borned when this song was fame-mouse, and mostly could Mum Please. Shut. Up!

April 8
You will be very actual pleased to know that the Fer-eee cross the Mer-zeee peoples are cleverer than Mum, and funnily a-very-nuff, they has either worked out that people might hassociate their fe-reee with the fame-mouse song, or they might be very quite sick of womens-of-a-certain-age-who-have-got-no-shame bawling that blinking song on their boat. They had hobviously

decided that playing the song over the tannoy system on the boat was very preferababble to listening to Mum trying to sing it.

I fink Mum was a bit actual disappointed. She did singing along with the tannoy, but said it wasn't the same as her doing it solo into the wind. She did do her bestest-best, and got lots of roly hiballs and disappearing hibrows from the other passengers, so I fink she was just about satisfied with her actual efforts.

Me and the previously ginger one did go and stand by the railings, and look at the booful buildings and the churning river. And also pretend that Mum didn't actual belong to us.

April 10

Dear Our Friends in the North

Fank oo for having me to stay at your actual house when I did come to Liverpool. I did realise I would be meeting noo hoomans, but noboddedy tolded me about Finn and Kelpie. I fink it would have binned actual quite very helpful for me to know this fing, because then I would have been a lot more hexcited about coming to see you.

I do know that I is a luffly boykin, but fink you did actual spoiling me by arranging for there to be 24 hour Rabbit Fee-at-her in the garden. Me and the rabbits did exerlent making friends and kisses. I could have watched all day. Finking about it, I probababbly did.

Fank oo for the trip to Liverpool, but it does seem that you has gotted Mum's hobsession with me trying out every single bitta public transport ever actual hinvented. So I can now add a Fer-eee cross the Mer-zeee, Liverpuddlian trains, and a teeny-weeny lift to the list of Fings Wot I Has Binned On. Or in. Or on top of. I liked them all, but for future actual reference, I aren't going in a small lift with a strange man. Or his bag. I do like hexclusive lifts with just me and my actual friends. And no strange bags.

I are sorry about jumpering on and lying on your sofas, but where I do come from, dogs do live on them; I fink it is The Law. Also, we is allowed on beds. So when you go into your spare room to see wot Mum has lefted behind, the spare bed might be a bit actual hairier than you is actual hexpecting.

Please give Kelpie some quite very respectful and gentle strokes from me, and say I is ever, hever so himpressed at her running round the park on her 16-year-old legs. Mum is gobsmacked at how fit and hagile she is at her age. I are sorry if I has teached Finn some new bad nabits, and also very failed to stop his one bad nabbit of finking all dogs be cushions-for-bonkering.

I hope Finn is not actual missing me too much. I did fink he was a quite actual fabumazing host, and I does hope I will be able to show him my part of the world very actual soon. You had better warn him about how far away it is, though. In case he finks we live on the moon.

From your luffly boykin

Worzel Wooface

Pee-Ess - please also do tellering Finn that at my house we has plenty of cushions-for-bonkering, so he won't need to do that fing to me.

April 12

Whilst I have been snoozing away my Liverpool hadventures, Mum has binned actual plotting and horganising more Progress. We will be gettering some noo fences in the garden, to replace the ones wot is held together by string and positive finking and some bits of trellis, and also a Noo Bit. The Noo Bit has not been discussed or hagreed with me so I are quite very hassuming I will not be pleased. Mum has done giving me some determined looks when the Noo Bit has been talked about, and there's been a few 'poor-Worzel' comments from Dad. You will be pleased to actual know I are already planning my Complaint to the Management.

April 13

Sometimes, it is not until you does go away that you can see some differences. When you do see things every actual day, it is hard to spot small changes, cos they sort of happen while you is not lookering, but this morning, I did go round to visit my noo neighbours, and I can tell you that my boot camp with Maisie, Kite's big sister, is quite actual working. Slowly, this is actual true, but it is working.

Do you remember me telling you about Maisie, wot was rescued from a biscuit tin? Well, her famberly have been working quite very hard at getting her slimmer, and I has binned doing my actual bit by being the bestest Boot Camp Commander in the whole wide world. Wot mainly hinvolves me doing running around and hinsisting that Maisie joins in. Usually, Maisie does take two steps, and then need to do careful lying down on the ground looking for her puff wot she is all out of ... but today she did running and playing and is getting a bit actual happier. And smaller.

April 15

The Progress in the garden isn't small and hunnoticeababble. It is fuge and loud and a blinking disaster as far as I are concerned. Two ignormous men have arrived to put up new fences and take down a quite actual annoying tree wot accidentally planted itself under one of our windows. I aren't sure how it managed to plant itself there, and I do fink that Mum might have hadded more to do with it than she is hadmitting to, but Dad says it's best not to point out Mum's gardening disasters or he could find himself In. Charge. of Gardening. And without any dinner. The ignormous men had lots of noisy tools, but I was a exerlent boykin and did exerlent supervising and not being worried a-very-tall. Both the ignormous men did like me. So I did like them.

All the trellis and the fences is mainly very okay, but the Noo Bit I do fink has binned designed to specially Ruin. My. Life. The Noo Bit is going to separate the side of the garden from the front garden, and the Hole Point is to stop Worzel Wooface from going into the front garden so that Mum can grow fings in there without me diggering them all up. This is missing the point.

The reason I do diggering up Mum's plants is cos there's a mole under them. I aren't trying to dig up the plants (well, not since Mum moved the really, wheely annoying ones that had spikes on), I are trying to dig up the mole.

The Noo Bit will be too actual high for me to get over, and too very low for me to get hunder. There is no gate there yet, though, so I has got a week to find that mole before he is locked away from me for ever. And hever. Or until Mum forgets to shut the front door.

April 16

My noo neighbours is very quite kind I do fink. After I has dunned running around like a loony doing Maisie Boot Camp, I do get very quite firsty. And I do need a drink. So I are quite actual pleased with the water-bowl-wot-honly-Worzel-can-reach that they has hinstalled in their garden. None of their dogs can reach it cos they is too actual short, and they has to do queuing up at the water bowl on the ground. But because I are special and a luffly boykin, I does get my quite very own actual one.

Mum says I are a plonker and it's really a bird bath, but I do find this quite actual hard to believe. Birds do not like baths. One felled in our bath once, and it had complete hystericals until Dad fished it out, and Gandhi got shutted out of the house and definitely put in the doghouse. So I can't see why a birdy would choose to do getting into a bath by its own actual choice.

The Dog-Hexpert will probababbly have hystericals if someboddedy does tell her about my noo waterbowl, cos she worries about everyfing, and wot else coulda been drinking the water. Well, I can tell her ... nuffink. Cos there's no water lefted in it now. I has drunked it all.

April 18

Well, it has tooked me quite a few very actual weeks, but I has FINALLY managed to add some new fings to my hooman biffer training.

Now, after I do biff, Mum and Dad do know that if I do biffing again and again, I want to do play. If I do biff and then walk away and turn around in circles, it means 'follow me, please.' And then I do more circles round and round, and take them to the fing I would actual like. Sometimes that is the door, cos I want to go for weesandpoos. Other times, I do take them to the kitchen cos it is time for my dinner.

I fink my famberly is quite actual clever to have gotted these three tricks horganised. But I would like to do pointing out that I has learned loads and loads more tricks than they has got, so I hope there will be no more talk of me being not-as-clever-as-all-very-that. And definitely no more hinsistering I do 'sit' before I is letted off my lead.

April 19

We had to call out a Hemergency Vet to our house last night, as Gipsy wasn't well a-very-tall. She was breathing through her mouth, wot for cats is a very bad sign, and we fort the Time Had Come.

After the vet camed to check her over, and after Gipsy had refused to stand still to be hexamined, then finally hagreed to have an hinjection but then changed her mind and decided instead to attack the vet with her claws, we all decided that there is probababbly still some more life lefted in Gipsy.

And then having made her forts and feelings perfickly clear, she then shot off out through the cat flap to get on with that life, leaving Mum feeling redickerless and the vet hunting for a plaster.

April 20

Kite's Mum says that she is mostly pleased that me and Kit have becomed friends, but wishes I hadn't taught Kite how to do Bitey-Facey. She says she can cope with the noises and the kerfuffling, but is quite actual shocked and halmost very frightened of the faces Kite does make.

I does not make these faces as nearly as actual well as Kite. I fink my hibrows and furry chin do ruin the himpact of pretendering-to-be-a-scary-Worzel-Wooface. Kite does not have a furry chin, and she does not have hibrows, so all of her I-are-going-to-eat-you-Worzel face is on display. She looks like the cartoon wot they use on the rabies posters, her Mum says, and I-wouldn't-want-to-meet-her-on-a-dark-night.

Well, I can tell Kite's Mum she has got nuffink to very actual worry about, as Kite honly makes that face for my very benefit. Anyboddedy who meets Kite on a dark night just gets wags and licks, even when they can't get no answer at Kite's front door, give up and just wander in and yell. Like wot Mum did last night. Kite would be a very rubbish wouldn't-want-to-meet-her-on-a-dark-night protectorer, unless it was me wanting to play bitey-facey.

April 21

It is Easter in Southwold, and also the rest of the world, Dad says. Easter isn't hexclusive to Southwold, which I do fink is a very good actual fing, cos if they honly did Easter in Southwold, a hawful lot of peoples would be quite actual disappointed.

Now it's Easter, there is nowhere to park your car, apart from Beccles, but that's about 12 miles away, and not wot you'd call convenient. The other actual halternative seems to be parking in the middle of the road, which a lot of the visitors fink is a totally hacceptababble place to leave their car whilst they nip into a shop. It isn't, but as they mostly come from London, I don't fink they hunderstand 'get-out-of-the-blinking-road-you-plonker.' Or how to drive the ignormous lorry-type cars they all hinsist on having, wot take up 17 car parking spaces each!

April 22

In winter, all the children I do meet have little names like Sam and Bella and Max. Once it gets to Easter, though, and then all the actual way through the summer, all the children suddenly get really, wheely long, complercated names, like Jupiter or Agamemnon. And they find it much harderer to do Being Good, so all you do hear is 'Chrysanthemum, do be careful, darling.' and 'Stradivarius, you'll never eat all of that, now put it down,' and stuff like that in really loud voices. Cos I fink they all go deaf as well.

Mum says they is Tourists: hoomans wot come to visit booful places and see how quickly they can hannoy everyboddedy who does live there all the time. Hexcept they has to pretend not to be hirritated cos they has got businesses to run, and fings to sell in the shops that they want the Tourists to buy. Mostly, the local people do remember this fing, but actual sometimes someboddedy will get quite very fed up, and moan that now they have to start queueing for fishnchips the night before they do want it, cos of all the Tourists. That person is usually Dad cos he doesn't want to be at the back of the queue listening to Tourists yackering on and on about house prices, and there being no halloumi or weird mushrooms in the Co-op, and that the World. Will. End.

Dad is meant to be bringing home fishnchips for tea tonight. If he can get parked ...

April 24

The gate for the Noo Bit in the garden has binned put on, so now I are never going to be able to snooze in the sunshine or smell the booful flowers or catch the mole ever, hever. I may be hexaggerating about that a ickle bit, but this is my Complaint to the Management, and I want it to be actual noted.

Mum says I are hexaggerating a lot, not just a ickle bit, as I *will* be allowed to do most of these fings ... but honly when she is out there with me. Apart from the mole-catching. Mole-catching is off the Nagenda cos the front garden is now Mum's special place for growing plants. If I want to dig, I can do it in the side garden, wot is quite very secure, and is now called Worzel's Garden!

April 26

When Mum wented out this evening and Dad was In. Charge. he forgotted about the Noo Bit and the Worzel-isn't-allowed-in-the-front-garden-to-dig rules, and let me out there. Dad isn't left In. Charge. very often, so I did decide I had to have one last manic try at catching the mole.

Dad's in the doghouse. And the mole seems to have disappeared ...

April 27

The fuge ginger boyman has comed back from Universally for a few days. I has binned trying out my biffer on him but he isn't himpressed. I've had to use my super-sized biffer on him to get any reaction. (To be honest, I don't have a super-sized biffer: I just stand on top of him when he is asleep on the sofa and woof at him until he wakes up. And then jump up and down a bit. And then do some more biffering until he stands up and lets me outside.)

Happarently, I are a Noaf, and really, wheely need to do lookering where I are putting my feet. The fuge ginger boyman tried to do Complaints to the Management, but Mum says the fuge ginger boyman is as bad as me, and at least I honly put my feet in the wrong place. Unlike him, who can't seem to put his boddedy in the right place. Like in a bed at night, rather than sprawled all over the sofa.

April 28

I aren't very happy tonight. I've had a yell and a sulk, and I heven had to have a very shaky lie-down on my actual own.

KITE BIT MY BUM! She did! And it hurted me a lot and a lot. Okay, well, an ickle bit, but I wasn't hexpecting it, and now I are a Deeply. Hoffended. Worzel Wooface.

Mum says Kite didn't mean to, and it was a haccident, and perhaps me and Kite shouldn't do playing quite so actual roughly with each other.

And maybe now we'll calm down, and watch where each other's teeth are. And bums ...

April 30

I has decided to forgive Kite for the hole bum-biting hincident. She did not actual mean to do it, we has had several weeks of playing nicely together, and I are pretty actual sure she is not planning to do it again.

After it happened, Kite did lots of slumping her shoulders and very looking away, then once I had finished yellering and sulking and pretendering I would never be able to actual walk again, she did careful happroaching me with some low gentle tail wag, trying-very-hard-not-to-look at me. It was all quite actual hobvious she was trying to make things peaceful and back-to-normal again, hespecially when she started to lick her lips and yawn. The best hinterpretation that the hoomans could come up with about this was "Aww, Kite's trying to say sorry, isn't that sweet?" when it was a lot more so-fisty-cated than that.

Today, when I did go to play with Kite, we did just carry on as if nuffink had happened. I did exerlent trusting Kite, and we did play our bitey-facey games like we always do. I fink Kite's Mum is a bit disappointed, as she did hope that Kite might decide it wasn't a nice game for booful Labradors girls. Kite isn't one of them sort of girls, fank goodness, and is back to doing her pretend vicious, wouldn't-want-to-meet-her-on-a-dark-night faces.

Visit Hubble and Hattie on the web: www.hubbleandhattie.com
www.hubbleandhattie.blogspot.co.uk • Details of all books • Special offers
• Newsletter • New book news

May

May 1

Happy May Day! I fink spring has finally sprunged!

We is quite very actual lucky where we do live. As well as having a beach and fields and our bee-strip, we has woods. We don't go in the woods in the autumn or winter, though, as Mum is always worrying about those strange bugs and illnesses that seem to live in woods. And trees landing on our heads. But in spring she can't resist because of the bluebells. There is billions of them: a hendless blue carpet of booful spring flowers. They is magnificent. They is stunning and, according to Mum, we is blessed.

Mum might feel blessed but I doesn't. As far as I are concerned they is a blinking nuisance. For a start, colours isn't something I are any good at. And just like the garden, I are expected to keep to the paths when the bluebells are out, and not do stomping all over them, and especially not do rolling about on top of them because a fox has dunned pooing under them. Mum shrieks then, and says I are spoiling the booful flowers, and geroff-you-great-heathen.

But the worstest fing is the photoing. Mum's gotted it into her head that she wants a photo of her booful Worzel Wooface, HAND the bluebells in the same picture. It's not going well. For a very actual start, I has got to do 'sit' and 'wait,' wot I are quite actual not happy about, so instead of a booful Worzel Wooface, I pull my usual please-don't-make-me-do-sit face that is quite actual shamed with a side order of hagony. I find this face is the quickest way to stop the sit stuff, and it usually works quite very well, unless the camera is with us. When the camera is out, Mum gets quite actual determined, and my shamed and hagonised face just means I have to do sit for even actual longer.

Second of-actual-all, Mum wants to get the perfick photo, but she really doesn't know how to get the bluebells blurry in the background, and Worzel Wooface in focus at the front, so we has to keep doing it again and again. And cos I wriggle off and stomp all over the bluebells, I spoil them, and so we has to go to a different part of the wood and try again (and again and again and again). At our current rate of success, there aren't going to be an actual lot of standing up bluebells lefted in the woods soon ...

And third of-blinking-all, the previously ginger one is fabumazing at doing photoing, HAND she can get the bluebells all blurry in the background and everyfing. So she ends up with the perfick actual photo and Mum does not. So she sulks.

May 3

****FUGE NEWS****

Mum's got me a noo mole! Just for me, and she's put it in Worzel's garden!

In quite very actual fact, I might have gotted that a bit wrong. Dad reckons that even though Mum is almost completely actual bonkers-crazy, and does love me to the moon and back, he's almost certain my mad diggering a couple of nights ago just made the mole scrabble under the fence out of the front garden, and into my bit of garden!

Mum's worrying now cos she didn't want to trap the mole anywhere, and doesn't want me to do killing it, wot would be cruel and wrong: they might make molehills, but they is dorable and a living fing.

Dad has a special face he uses when Mum says fings like this. It's called his George O'King face. Well, that's wot I call it, cos it's wot he always says when he uses it. Happarently, Dad reckons I has had two years of trying to catch the mole, and I has failed dismally so far, and anyway, if Worzel gets too close again, the mole will just scrabble back under the fence into Mum's garden.

Me and Mum don't have any special faces for Dad tonight, but, then again, Dad hasn't got any dinner ...

May 4

I are a proper growed up big boykin, Sally-the-Vet says. I did just have my booster hinjections and did not need to have any sleepy meds or nuffink. I tooked it like a, well, Mum, really, and definitely not like a manly Dad. Dad hides hunder the duvet with his phone, and cancels Opital appointments cos he's a scaredy boykin when he has to be fiddled with. Mum has to phone up and huncancel them and drag him there under Frets of No Ketchup, which is about the worstest fing that can happen to a Dad, happarently.

Anyway, I did have all my hinjections without having hysericals, and I are mainly feeling quite actual proud of my very self. So if it is all very quite okay with everyboddedy else, I are going to go and find my Dad. .. and hide under the duvet with him till I has forgotted about it.

May 5

I aren't wot you could call a silly dog. Or daft. Or goofy. Wot I are is quite often an anxious boykin who worries too very much about everyfing. But even though I will be three in June, I are still learning noo fings.

Just recently, I has gotted to know that chasing a ball (and very actual not giving it back) is somefing that is fun. I used to sometimes do chasing a ball for Mum, but honly really to make her happy, and I always had a bit of a worried look on my face, as I wasn't actual sure I was doing the right fing, or if I would get tolded off.

But recently, I are gettering more and more confident about it. And will look for a ball to play with and be nearly hexcited about it, hespecially when I are with doggy friends. I are quite very glad Mum didn't give up trying to get me to like toys, even though it has tooked nearly two years.

May 6

I does not know wot the Heckington Stanley has happened to Maisie. When I did see her this morning she was being most actual quite per-cooly-ah. She wants me to be her Noo Best Friend. I mean, her really, wheely Noo Best Friend. Maisie isn't usually this actual pleased to see me. Mostly, she does want to shout at me when I go into her garden, and then she gives me a sniff before deciding that watching me and Kite zooming around is too quite hexhausting.

So she claps under the table. But now she won't stop waddling after me in a very quite fusey-tastic determined actual way. It is most confuddling ...

May 9

Maisie is still being redickerless and, to be actual honest, is spoiling my bitey-facey ball chasing games with Kite. According to Maisie, Kite is Not. Allowed. to be my friend any-actual-more, and I are honly very allowed to be friends with Maisie. She does not want to do bitey-facey or play chase; she wants to do, well, other fings, wot is quite actual not suitable for plite comp-knee. I are quite actual hembarrassed to say that Maisie doesn't want to be a chubby Labrador anymore; she wants to be a cushion-for-bonkering.

I has met a few dogs wot did want to bonk cushions, and a few wot have wanted me to *pretend* to be a cushion, so they could bonk me, but I has never, hever metted a dog wot wanted to be the cushion wot gets bonked. And currently, I does not want to meet Maisie, and I certainly does NOT WANT TO BONK HER.

May 10

The previously ginger one has had some hexciting actual noos. She has binned asked to write a noospaper collie-um-num for our local paper, all about wot it is like living with a Men-Tall Elf, and to help Fight the Stigma.

She's having a bit of a panic about being nineteen, and not that actual good at writing like a proper grown up. The Head Hitter has tolded her not to worry about that as he wants lots of different actual 'voices' in his noospaper, and not everyboddedy sounding like they is 45, and went to Universally to study Ingerlish.

I does hope the previously ginger one does sticking to her nineteen-year-old, just-about-normal voice. And not let the voices in her head wot tell her she's rubbish and useless be in the noospaper.

Them ones should shut right-blinking-up as far as I are concerned.

May 11

I has binned tolded wot Maisie's problem is and I does not like it. Maisie is Hin. Season. Being Hin. Season. is when luffly girlkin doggies do decide that they want to make puppies. I do not fink it would be a very good hidea for Maisie to make puppies. I does not want to cast nasturtiums about Maisie, but she is too waddling chubby to keep up with puppies, and she is nearly a senior doggy, HAND she hasn't dunned making puppies before. All in-very-all, it would be A Disaster.

I does definitely not want to be the doggy wot makes puppies with Maisie, heven though she is quite very HINSISTING I do try. And the honly reason she is hinsisting I do try is cos she's honly got me and Kes to choose from, and Kes is far too busy horganising the hedge into his perfick scratching post, and also would probabbably fall over if he did try to do bonkering. I are Obson's Choice, and a bad Obson's Choice, to be quite actual honest cos I couldn't make puppies with Maisie even if I *did* want to. And that is cos I don't have all the bits wot are needed to do that fing.

THREE QUITE ✓*very* ~~actual~~ CHEERS FOR **Worzel Wooface**

Wot does give me another blinking reason for why Maisie should not do having puppies: they might get her brains. Wot is fick, hobviously. There is too many actual puppies in the world as it is; the last fing we need is puppies wot are as fick as Maisie.

May 12

Today is National Limerick Day, and I has struggled big-time-badly with this actual challenge. This is the best I could quite very do –

> There once was a doggy called Woo
> Also Worzel and Wooface, it's true
> But Worzel don't rhyme
> With nowt I can find
> And I struggled with Wooface, too

I aren't sure about the Nowt word neither. Nowt is a North-ish word for nuffink, but I aren't North-ish like Finn and Kelpie so it feels like cheating. Dad says it's quite actual okay: when you is poem-doing and desperate for a word wot fits, you can use anyfing you like. It's called Hartistic Lie Sense, and at least I'm not blithering on about being lonely as a cloud ...

May 13

This weekend I has been cunningly havoiding Maisie and her wanting-to-be-a-cushion-for-bonkering cos Lola has been staying with us. Lola does not fink she is a cushion, and she does not want any puppies, I are actual relieved to say, so we has had some fuge bitey-facey games, and we went to the marshes for playtimes. There has binned lots of rain, so all the soggy bits on the marshes were filled up.

Whilst we was at the marshes I did notice billions of molehills in-between the cowpoos and bird droppings and clouds of crows wot are living there at the moment. It was all nearly very too actual much for a luffly boykin to take in, and I didn't know wot to do actual first.

I knew wot to do last, though, and when it was time to go home, I did exerlent hignoring of Mum. My I-don't-wanna-go-home plan worked perfickly until Lola, being a Labrador wot can't say no to a bitta cheese, did let the hole side down, and went sneaking back to Mum when I was not lookering.

I are quite Orrified and Hoffended to say that I has a Bad Mum, a really, wheely Bad and Orrid Mum who did do habandoning me. She marched off towards the gate, giving Lola MY cheese, and did hignore me and my I-don't-wanna-go-home fandangos. So I had to do rushing after them to make sure I was not forgotted.

I do not fink it is my job to make sure I are remembered about, but when I caughted up with Mum and saw the smug look on her blinking face, I do not fink I had been forgotted about at-very-all.

I fink I had been Not Forgotted About a quite actual lot. And I was tricked. Again.

May 14

The Head Hitter is very actual pleased with the previously ginger one's first go at a collie-um-num, and he has hardly changed nuffink wot she wrote. Apart from the hexclamation marks, cos you honly need one. Not four. No actual matter how blinking cross you is about somefing.

How-very-ever, cos this is not a noosпаper collie-um-num, and I aren't nineteen, and I *are* flipping cross, I are going to use as many hexclamation marks as I blinking well like!!!!

I Are Cross. With. Dad and I are Not Speaking to Him!!!!

This morning Dad did get up and go to work, and did decide to take Lola with him. But not me. He says this is cos I don't like waiting in the hoffice when he has to go and do outside work, but I do fink it is cos Lola follows him around like a proper hobedient dog wot makes him look clever. It doesn't. She just follows anyboddedy with food in their pockets, and I are a bit more hinterested in going to the marshes. And the pub. And basically bogging off and showing off my recall-as-good-as-a-freezer.

When I did realise that Dad was taking Lola and Not. Me. I did some concerned woofing. And then some hoffended jumpering up and down on the bed trying to see out the window. And then some more Houtraged at Dad Complaints to the Management.

Hunfortunately, the Management was still asleep and trying to have a sneaky lie-in. I might have dunned jumpering up and down on her boobies a bit as-very-well, wot is not allowed and really hurts-you-great-nana, and not a great way to begin a day, Mum says.

May 15

Tomorrow I are going to the National Pet Show in London, and I are really, wheely looking forward to it. I are going to be on the Hounds First show stand to be a exerlent hexample of a rescue Sighthound. Mum says there is a nice safe private crate bit at the back of the stand for when I have had a-very-nuff of being a exerlent hexample, and do want to be hignored and on my own, so I fink it will be very okay.

But there is going to be Somefing Sig-niffy-cant happening as actual well. After two years, I are going to meet Rachel again. She was my first mummy from Hounds First, and she is going to bring my old pal, FeeBee, her Whippet, to see me. I are quite actual hexcited and hinterested in all of this, but a ickle bit nervous as well. I wonder if she will fink I are a luffly boykin, and whether Mum has dunned a good job with me. Rachel is one of them Dog-Hexperts as very well. I has got a horribibble feeling she's going to want me to do sit and stuff like that, and I are rubbish at that kind of fing.

Dad says Rachel and Mum are going to do crying. I don't fink I want to be part of that bit. Private sobbing into my fur is just about very hacceptababble, like after a soppy film or when there is bad news. Public red-wobbly-face-doing when I is on a lead and can't run away bravely and pretendering I Aren't. With. Them is very Not. On.

And Dad isn't helping cos he's managed to come up with somefing himportant he's got to do with a customer at the harbour, so he's wriggled out of coming, and is quietly and very actual not-in-front-of-Mum punching the air in the kitchen and whispering yes! yes! yes!

May 16

All roads should be made of the same fing, I do fink. It should be a actual proper law cos I do find it quite very confuddling, and need to have-a-panic-in-case-the-world-is-going-to-end when roads change from concrete to tarmac. And bumps and roadworks and radios in other cars should be banned for the same reason. Apart from these fings, I did enjoy my trip in the car to the National Pet Show.

When we gotted to the Pet Show we did do going to find the stand wot we was to be on. And then I did see FeeBee and Rachel. My Wee-union with them was quite actual odd. You know when you sees someboddedy wot you has not seen in a long time? And you fink you remember them but you can't work out where from, or if they is just a fame-mouse person off the telly, so maybe you don't actual *know* them, and you is probababbly going to make a hidiot of yourself? Well, it was like that. I knewed I did know them from somewhere, but I could not do working it out. And cos some of my memories from a long time ago are not good memories, I did do quite a very lot of pretendering I did not know them until I worked it all out. Fortunately, FeeBee did not care wot I did remember, and decided we was going to be friends again straightaway.

I was quite actual glad FeeBee was at the show, as she was a fabumazing hinfluence, and helped with showing me some ropes about How to Be a Exerlent Hexample of a Rescue Sighthound, and did lots of saying 'hello luffly peoples' to the visitors to the stand, wot gave me confidence to do the same fing. And when I did get too tired I wented and had a quiet time in the safe space at the back of the stand, and let FeeBee carry on with lying on the carpet and face-lickering.

There was also another doggy there called Archie, who did decide he did not want to play with me. So I did havoid doing playing with him. And then he did decide he did want to play with me, after-very-all, and did biffering me with his nose. I fink we camed to a hunderstanding in the end, but Archie is about as complercated as me in the head-finking department ... I do fink.

FeeBee, on the other actual paw, is not complercated in the head-finking department. She was just happy to be nearly stooded on, and stroked and played with and cuddled and ... I fink there will be a lot and a lot of henquiries to Hounds First for a Whippet-just-like-FeeBee, cos she was perfick at all the meeting and greeting. I did my bestest but it was actual noisy, and I did need some chill-out time. But I was quite actual good at moving away when I felted under pressure, and did not do anyfing to let myself down.

There was a lady on our stand called Tina and, at one point, I did decide that she and me needed to go for a walk. So I just walked off with her and took her outside, cos Mum was too busy yacking and being helpful. That did surprise everyboddedy, but Mum was quite actual pleased I did trust Tina a-very-nuff to do that fing.

Finally, after about five hours, I did decide it didn't matter if I remembered Rachel or not as she was quite very okay and we would be friends. And I are pleased to say there was no crying, wot I did see. But I did hear Rachel and Mum talking about me looking wonderful, and it was very actual hobvious that I was living in the right forever home.

I did do lettering myself down quite actual badly at one point, though, and I do blame Fish4Dogs for that fing. I did sit. Without being asked. Wot is wrong and Orrendous, and I must try to do much betterer at remembering in actual future. Hespecially as I now seem to have learnt 'down' without meaning to, neither. Well, you woulda dunned as well, if you'd been on that car journey with the roadworks and the radios and the bumps, and me standing up to check that there wasn't a hapocolips going on. And Mum not being able to see out of the back window. I've binned conned into another hobedience fing, and I are Not. Happy.

All-in-very-all, it was a long but fabumazing day. I are completely hexhausted now, and planning to sleep for a week.

May 17

There were lots of childrens at the National Pet Show. I does not live with childrens, so I honly do hoccasionally get to meet them. I are still trying to work childrens out, but my old pal, FeeBee, who I did also meet at the National Pet Show, does live with SIX childrens, and she did giving me lots of advices.

****FORTS ABOUT CHILDRENS****

- Childrens are young hoomans, like puppies are young dogs. Unlike puppies, though, they do not like it if you lick their bums
- In-very-general, children should be sniffed before being heard
- Childrens come in different shapes and sizes. They all make a lot of noise
- Childrens can be very actual scary, so it is best to make sure you always have a hadult with you in case they do something weird
- If there is lots of childrens about or there is one that is being too very noisy and flappy, go to your bed. Most childrens do hunderstand that when you is in your bed, you want to be lefted alone
- Most childrens like mud. And running. And playing. And they don't want to go inside when it's cold, neither
- Let sleeping childrens lie. Seriously. Woofing at the doorbell and waking childrens up will make you very actual un-pop-oo-la
- If you has the hoption of cuddling up to a childrens or a hadult, choose the childrens.
- In-very-general, they does not keep getting up to cook dinner or answer hemails
- The smaller a childrens is, the more hunpredictababble they is. Until they does become nearly hadults and find out about cider. Then all bets is off
- Dads is not childrens. Even if they does want to be

May 18

I had a strange hen-counter of the Collie kind on the way back from the National Pet Show. Mum had booked for us to stay at a hotel overnight, that was very quite nice for doggies, and had a fabumazing field next door for having a mooch. The peoples on the desk were all actual quite keen on dogs, wot was a good blinking job cos there was billions of dogs there. And, apart from me, they was all Border Collies.

You does probababbly know that I is a bit not-very-actual-hinterested in hobedience, and worried that the Dog-Hexpert will fink it is somefing I should do more often. I has discovered why I aren't very much good at it and, more himportantly, it isn't my fault. It's Mum's.

Mum can't do Magic wot is hobviously hessential for that hobedience stuff.

'Wait' said the Magic-woman-with-the-Collie. And the Collie did stand in the entrance of the van, doing 'wait' like she had binned glued to the ground. And then the Magic-woman camed over and did talking with Mum, and heven though I was there wagging my tail and wiggling my boddedy, making it quite actual clear that I would like the Collie to come and say hello-luffly-boykin, and sniff my gentleman bits, and all that kind of normal doggy stuff wot was very actual hexpected, the Collie did just do stand and wait. Magic, it was.

Either that, or the treats wot the lady did give to her fluffy Collie after she did wait for all that time was super-special, and nothing a-very-tall like the ones Mum has.

Mum tolded the peoples that we was on the way back from the National Pet Show, and the lady said that all the Collies were going to a somefing Ticket Show. I fink they must have been going to get tickets to Mars of somefing cos I are sure them Magic super-powers could be quite very useful to save the galaxy.

I was a luffly friendly boykin at the National Pet Show and I didn't do letting myself down a-very-tall, but I didn't do stand. Or wait. I mainly did snoozing and wagging my tail, cos I is not Magic, and neither is Mum. And I really, wheely don't want a ticket to Mars ...

May 19

Gran-the-dog-hexpert has tolded me wot a Ticket show is all about. Happarently, every year there is a fuge compertishon to find out who is the most hobedient dog in the country. And cos lots of peoples fink they has got a clever dog, they have mini-compertishons, and if you win one of these mini-compertishons, you win a Ticket to have a go in the finals.

Gran says I are a luffly boykin and I should not do worrying-my-pretty-head-about-it. I aren't sure whether I should be feeling relieved or quite very actual hinsulted now...

May 20

Can you quite very actual believe that Gandhi is two years old? It does not seem that very long ago that him and his brothers and sisters were ickle kittens wot keeped falling in my water bowl.

Gandhi wouldn't fit in my water bowl any-actual-more. In fact, he is starting to get almost as fat as Frank, and Mum does not hunderstand why he is gettering so ignormous. Frank has gotted an hexcuse; Frank was a starving stray wot eats for Ingerland, cos of wot happened to him when he was lost and starved. Gandhi has never binned a stray or starving; he's just a hogpig.

Mum reckons that maybe Grey-Cat-with-No-Tail has gived Gandhi some hideas about breaking and hentering, and will be having a word with the neighbours before he does hexplode.

May 24

Mum has dunned asking all the neighbours if they has binned having strange hen-counters of the Gandhi kind in their kitchens, and none of them has seen him. Or seen their cat food disappearing at a halarming rate.

So, unless they is very quite hunobservant, there is either some very quite fick and slow bunnies in the fields, or Gandhi is Up. To. Somefing.

May 25

I don't fink Gandhi is catching bunnies. For an actual start, he woulda bringed them home. And left bits lying around. There is a habsolutely disgustering bit in the middle of a bunny wot no cat will eat, and usually when there has binned bunny-bashing, we come downstairs and find it spatted out somewhere in the kitchen.

Mostly Dad manages to havoid stepping in it on the way to work. Unless it's binned spatted out next to the kettle, which would be quite actual hard for a hooman to stand on unless they was trying actual quite hard. And being weird.

Anyway, there has binned no spatted-out bits lefted around. And to be quite actual honest, Gandhi is so ignormous I don't fink he could catch a bunny at the moment: he's starting to find it quite very difficult to jump up onto the table, even.

May 26

It was Gypsy-the Foster-Fridge's Gotcha Day a few days ago, and I forgotted to say 'Happy Gotcha Day.' Usually, Mum is quite actual good at remembering to remind me about himportant dates, but she's binned busy with other fings. Like hinsisting Dad does Somefing. About. The. Ivy. wot Saint-Fred-who's-a-bit-deaf-fank-goodness, who lives next door to us, has been asking us to sort out for weeks. Not honly is the Ivy coming in our windows now, it's trying to get down our chimberly, and eating a-bit-deaf-fank-goodness-Saint-Fred's telly wires.

So Dad had to get out a ladder. And Go. Up. It. And be manly and not have hystericals, wot he would have preferred to do, I reckon. I do find this quite actual strange cos we have pictures of Dad climbing up masts with no bother a-very-tall. He says that when he goes up a mast, he is attached to a bitta rope and it's different. I fink it's just different cos it's boat work, and climbing up ladders to get the ivy out the chimberly is too actual near to gardening, wot Dad is hallergic to, and it makes him go all faint and wobbly, and quite actual grumpy.

Dad's also hallergic to Dee-Aye-Why as well. Every time Dee-Aye-Why comes up in the conversation, Dad gets the runs, big-time-badly. He runs off down the harbour to play on his boat, and stays there until Mum forgets about dishwasher cover panels and broken garden chairs.

But now Dad's in fuge troubles. cos now that Mum has remembered about Gypsy-the-Foster-Fridge's Gotcha Day, she has remembered that it is now TWO YEARS since she and I did busting the banisters, and Dad still hasn't fixed them.

Mum says the banisters are now Right. At. The. Top of Dad's list of Fings to Do ... if he ever comes home from the harbour.

May 27

We has had a cunning plan about Gandhi and his hexpanding belly. Our friend, Auntie Kate, does have a tracker fing! She did use it for when one of her actual cats did keep bogging off and she needed to go and find him. He had a nabbit of getting stuck in peoples' sheds or jumpering in cars wot did make life quite very hexciting for Simba, but a bit stressful and tents for Aunty Kate.

Mum has decided she is going to stick the tracker collar fing on Gandhi to see where he is actual going, and find out why he is getting so blinking fat. I aren't too actual sure Gandhi does want his out-of-the-house-life being part of Mum's business, but she says that's tough; it *is* her business. Having one fat cat is a Miss Fortune; having two fat cats is careless. And hembarrassing.

May 28

I aren't sure if Gandhi fort he was helping Mum with her hinvestigations or not, but when Mum putted the collar on him for the first time this morning, Gandhi did shoot out of the cat flap like he was planning on hexploring the moon, and needed to get started straight a-very-way. Either that or he wasn't too actual himpressed about wearing a collar. Hunfortunately, he did not wait for Mum to attach the helectronic fing wot clips into the collar and does the tracking, so if he was trying to be helpful or show off with his aren't-I-clever-I-can-get-to-the-Moon mission, it will all be wasted. The tracker bit is still in the kitchen with Frank, and has gone about as far as he has. The most hexciting fing the tracker has managed to record today was getting flicked off the table by Frank's tail, and it hasn't moved since. Neither has Frank.

May 29

Gandhi has comed back from the moon and it's a disaster, cos he's comed back without the collar. You is probababbly finking that losing a cat collar shouldn't count as a disaster, and normally you would be quite actual right, but the collar had a sneaky and himportant and complercated piece of plastic on it wot the tracker fits inside. And when Mum hemailed the peoples that make the tracker, they doesn't have no spare bitsa plastic to give to her. The honly fing they could actual suggest was buying another tracker, but Dad has said very actual No-with-Knobs-On, partly cos they cost a flipping fortune, and that is moneys he could spend on his boat, but also On Princey-Ball. And a lot of words beginning with B about daylight robber-dobbery, and he's not paying fifty pounds for a penny piece of plastic. And he'll bring some gaffer tape back from work tomorrow.

Mum did a quite actual lot of wailing about the tracker, and the penny piece of plastic belonging to Auntie Kate, and how was she going to tell her that she'd lost the collar, and you-can't-just-borrow-someone's-stuff-and-only-give-them-half-of-it-back.

Dad says Aunty Kate will understand: she's got a boat.

May 30

The mystery of Gandhi's Belly has binned actual solved. Solved is not the same as sorted out, though: sorting it is going to be complercated and delly-cate, and quite very possibibbly himpossibibble. And it wasn't the tracker wot solved the mystery, it was actual me!

On the way back from our walk yesterday, I did spot Gandhi in someboddedy's garden, and did my looking-very-hinterested-to-see-if-he-would-run ear pricking that got Mum's attention. But Gandhi was not hinterested in running a-very-tall, cos he had his head in a plate of cat food wot a kind senior lady, Mum has now discovered, puts out for the hedgehogs.

Currently, we has got no solutions to this problem. We can't stop Gandhi pinching the food, and we don't want to spoil the senior lady's hedgehog feeding cos she loves her himportant work. After I did exerlent lookering hinterested and ear-pricking, I did kindly offer to scrabble hunder the hedge and drag Mum with me to stop Gandhi eating the hedgehog food, and also finish off any leftovers, but happarently this wasn't a solution, neither. And getting dragged hunder a hedge by a hidiot-donkey-of-a-Lurcher is just as hembarrassing as having two fat cats.

May 31

Mum and Dad wented out on Friday night to celebrate living in the same house, and putting up with Mum leaving the larder door open, and Dad piling everyfing he owns in front of Uff-the-Confuser instead of putting it away. And still quite liking it, apparently, even after twelve years.

Whilst they was out and I was lefted Home Alone, I committed a Cardinal Sin, Dad says. Mum says we're lucky it wasn't a cardinal hexplosion – a great big, oozy, sticky, bright scarlet, vermillion-and-cardinal disaster – and could Dad please remember to not leave the ketchup bottle on the sitting room floor, and how about putting it away next time?

I honly chewed the top of it. But then the lid pinged off, and I fink that's probababbly wot saved us from a ketchup-coloured hapocalips. Cos when that happened I did decide to bravely run away up the stairs, and founded that Mum's shoe-wot-hurts-her-foot-so-that's-not-such-a-bad-thing-really was more to my taste. And shoes do not have lids wot ping off hunexpectedly and disappear down the back of the sofa. At least that's where Dad did find it, after a bit of huffing and puffing, and words beginning with B.

The good news is that, since this hincident, the ketchup has tooked up living in front of Uff-the-Confuser on Dad's desk, along with everyfing else he finks is 'his,' and has becomed part of the ignormous anti-cat, don't-sit-in-front-of-my-screen-or-people-will-die barrier. Where it is actual very safe from any more lid-pinging, Worzel-scaring hincidents. Mum's asked Dad wot's wrong with him putting it away in the larder, but Dad says that would be pointless, cos Mum never shuts the door ...

JUNE

June 1

I've got a noo facilerty! Mum says it's grass. It isn't. It quite very actual definitely Is. Not. Grass and I has hevidence. I does know it is Not Grass because I did sittering on it. I honly sit on carpet, or if I is out on a walk, and Mum says I'm-not-lettering-you-off-your-lead-unless-you-do-sit when she is showing off to her friends. So it is definitely Not. Grass, it's carpet, and if I are struggling to get my head around this, then I does Dread. To. Fink. how the cats are going to do coping.

We've binned here before with this inside and outside stuff. When Dad broughted in a tree from outside one Crispmas, Mouse-the-cat gotted very quite confuddled and fort she could do weesandpoos behind it. And none of the hoomans fort this was very okay. So I are wondering why they fink it is a good hidea to do muddling indoors and outdoors again with this grassy carpet.

Mum isn't finking about that. She finks the noo grassy carpet looks fabumazing, and is much, much betterer than the muddy squashed flowers that were there before. It's too actual dark under the napple tree, she says, and anyfing that tries to grow there gets stomped on by a fuge Lurcher lookering out for his Dad to come home.

And all this might be quite very actual true but it is still carpet, it is still outside, and Mouse is still really, wheely fick as well as being lazy. I fink she's going to find this Not-Grass-really-carpet even harderer to get her brains around than me.

June 2

The previously ginger one's Men-Tall-Elf has binned causing her all sorts of troubles this week, and she's had to go into Opital again. Fortunately, now she is hofficially a hadult, there is an Opital quite near where we do live so Mum and Dad have both binned to visit her. I aren't allowed to visit but, happarently, it is like a really, wheely boring hotel wot always smells of cabbage and not chicken wings, so it doesn't sound like my kind of place.

The house is too quiet, though, and it feels like there is not a-very-nuff heartbeats here. Mum says I are helping her whilst she is missing the previously ginger one, just by being around and keeping everyboddedy in a routine. And wagging my tail when I see her.

> I can't fix all the problems
> I can't make 'em go away
> But sometimes me just being me
> Makes my famberly's day

The previously ginger one will come home in a few days, and, in the meantime, Dad says I could make his day by not sneaking into the previously

60

ginger one's bedroom and pinching all her crisps. I fink Dad might have quite very missed the point of my poem, to be actual honest ...

June 5

Mum and Kite's Mum are up to no good again. Last night they did have too many drinks of wine, and did decide to start moving around all the chairs and tables and plant pots in Kite's garden. And by the time they had finished drinking the wine, and giggling and dropping stuff, and telling me and Kite to go and play, and Maisie to move from under the table before she gotted somefing dropped on her head, they had created ... A Fuge Mess and a long list of jobs for Kite's Dad.

It all started when Mum went to find my paddling pool in the leantoo. The leantoo is a bit of half-built shed that hangs off the real shed in our garden. And the real shed is in a secret bit of our garden surrounded by a fence. Mum very, very quite oppy-mistically calls this bit of the garden The Courtyard. It used to be Dad's working-on-the-car-yard. It should still be Dad's working-on-the-car-yard, and Mum promised him that all the plants would be in pots, so that if he ever did need to work on the car, he could move all the pots to get the car in so he could do that fing. That was five years ago and Mum really, wheely did mean it at the time, but she forgotted that plants grow, and now most of the plants would need a crane to move them. So, it isn't Dad's working-on-the-car-yard anymore, it's Mum's Courtyard.

But the leantoo doesn't belong to Mum. It belongs to everyboddedy, and it's the place where my famberly puts fings they don't know where else to put, like the minidigger-in-bits wot Dad will fix one day, and the forgotted-about dumbbells that the fuge ginger boyman needed for about a week six years ago, and a roof rack for a car noboddedy can remember driving. And somehow, my paddling pool was hunder all of that stuff.

Mum did quite actual well getting the paddling pool out of the leantoo. And making sure all the fings that the paddling pool had stopped from crashing to the ground stayed where they was. It wasn't until she tried to get her-actual-self out of the leantoo that she realised that the fing stopping all the other stuff crashing to the ground was, well, her. And she was quite very actual stucked.

After Mum had got quite very hot and dripping trying to untangle herself, and also dunned frightening Mabel to actual very near death with all the crashing and yellering for Dad to come and rescue her, then panicking cos he was at work and wouldn't be home For. Hours, Mum remembered she had her phone in her pocket, and called Kite's Mum to very hexplain that she was stucked. And possibly about to end up deaded and buried under the minidigger, and if that fing happened, Dad didn't do it. Wot he couldn't have dunned cos he was not at home, and if he had binned, she'd have got him to get the somefing-beginning-with-B paddling pool out, and she wouldn't be in this sit-you-nation. But mainly, could Kite's Mum come and save her quite very actual quickerly? Like Now ...

You might be actual wondering why I did not do exerlent saving of Mum. Well, for a very start I are a dog. And I does know this will come as a fuge shock to some peoples, but Lassie wasn't real. Well, there was a dog and there was a film, and I are sure that all the dogs wot pretended to be Lassie were quite actual clever at the run-in-that-direction stuff, rather than the someboddedy-is-stucked-down-a-mineshaft-and it's-going-to-clapse-if-I-don't-woof-at-the-boy-with-the-blonde-hair (whose name noboddedy can remember) to get the man (whose name noboddedy can remember, neither) to rescue the someboddedy-down-the-mineshaft-who-was-diabetic-and-needed-hinsulin. Wot is also wrong cos if he was stucked down a mineshaft he'd need food, not hinsulin, but bring-a-banana isn't nearly so very hinteresting and hexciting as hinsulin.

So I did wot almost every other sensibibble and wise actual dog in the world would do if there was lots of scraping and heaving and words beginning-with-B coming out of a leantoo: I ranned away and hid on the big bed, and did exerlent pretending it wasn't happening.

I don't fink Mum did a terribibbly good job at hexplaining the hurgency of the hole leantoo-going-to-land-on-my-head to Kite's Mum.

"Ooooh," she said. "This is nice."

I shall not do repeating wot Mum said next, but actual quite hadmiring The Courtyard was not the pri-orry-tree, apparently. Once Mum was rescued, though, and had checked that Mabel hadn't died of art failure, she and Kite's Mum did both spend the time gettering their breaths back by having a good look around The Courtyard, and Kite's Mum decided that she would like a Courtyard as well. But not a leantoo. Or a minidigger.

June 6

After all the troubles gettering it out of the leantoo, my exerlent pretending the paddling pool does not hexist and very careful hignoring of it is hunacceptababble. Usually, Mum is quite actual patient about me hignoring stuff, and waiting for me to decide I like somefing ... but this is Ingerland. And in Ingerland, you can never be sure if the week of sunshine we is having will be the last one we ever get. She says I should do hurrying up and get in it before it's winter again.

June 7

My paddling pool is now at Kite's house. Kite is about as good at hignoring stuff as I are at ironing or playing the guitar, and it tooked her less than fifteen seconds to decide that my paddling pool was not to be hignored, and we should do playing with it.

There was a moment when I did have to do hiding at the bottom of the garden when Kite's mum did filling it up with the snaky-spitting-fing, but after that everyfing becamed very okay ... and then completely actual fabumazing. I are hoping that the weather stays sunny for a bit longerer so we can do more playing with the paddling pool at the weekend.

****FORTS ON SUMMER****

☀ Summer in Ingerland is supposed to warmer than winter in Ingerland, but quite very often isn't

☀ The hole point of summer is so hoomans can moan when it is raining, and also moan when it is too hot. I fink they mostly just like moaning

☀ And even though it has binned doing it for billions of years, they do get all hexcited that the days have got longerer. Like it's the first time it has ever happened

☀ In summer, some hoomans change colour. To red. It is bestest for luffly boykins to keep their feet away from the red bits. Hespecially their nails

☀ When hoomans say they is going to the beach on hot, sunny days, do not be actual fooled. You will really be parking on a motorway listening to traffic hupdates on the radio.

☀ If it's my paddling pool from May until October, it can't suddenly be OUR paddling pool for three weeks in August. Please get your own

☀ And don't even fink of washing your muddy gardening hands in it
Your swetty bits taste really, wheely nice

☀ I would like to snooze in the sunshine. I does not really want to be woked up every five minutes to check I aren't dying of thirst or heatstroke. If I is too hot, I will stand up and go inside. I aren't stoopid

☀ But doggies can't get up and go somewhere cooler if they is stucked in the car. Please remember that very himportant fing

June 8

We has binned hinvaded and it is Orrendous. Hole cuppateas are being made actual disgustering, which, according to everyboddedy here, is The. End. Of. The. World. My famberly's weapon of choice to defeat the hinvaders is somefing called Sorcers, honly they can't actual find any, and there is no actual point Mum lookering at me and saying 'where-did-I-put-them?' cos I does not know. I has binned living here for over two years now, and I has never seen a Sorcer. Dad says he's binned here over ten years and he hasn't, neither.

But Mum is actual convinced we has got some, somewhere. And they is the honly fing wot is going to stop the billions and billions of little flies that have hinvaded us overnight from getting in the cuppateas.

So, first fings actual first, Dad says: find the Sorcers, save the cuppateas, and then work out where all these blinking flies is coming from. Cos hobviously, noboddedy can go into battle without a cuppatea.

June 9

The previously ginger one has comed home from Opital, and she is feeling a bit betterer. Whilst she was in Opital she did remember that once-a-very-pon-a-time she was good at drawing. And she is doing lots of art, wot is helping her to stay positive about getting betterer, and stopping her finking that she would be better off doing dying.

None of us can really hunderstand why her Men-Tall-Elf would make her fink this fing. Dad has tried to remind her that she is fabumazing at drawing, and the Head Hitter is quite actual pleased with her collie-um-num, and lots of peoples do love her, and ... She says she knows all of this. Knows. All. Of. This. and it doesn't help: it just makes her feel quite actual guilty on top of everyfing else.

June 10

It is my himportant work to do gentle and cheerful tail-wags when I do see the previously ginger one. I are trying my bestest best to live up to my poem idea of 'me just being me,' and I do fink it is being very quite actual happreciated.

I are also working hard at remembering not to pinch any more of her crisps unless they is actual offered. Hespecially not the prawn cocktail ones. And also not pinch any more of her crisps unless they is actual offered. Hespecially not the prawn cocktail ones.

June 11

It is Mum's himportant work to stop asking the previously ginger one if she is okay every five minutes, cos it's driving her bonkers. A different kind of bonkers to the Men-Tall-Elf sort, and one that can be easily sorted if Mum would honly SHUT UP. And also make her a cuppatea, with lots of sugar, and preferababbly NO FLIES.

June 13

Kite's Dad says that Mum is a Bad Hinfluence. He has binned lumbered with building a Courtyard fing in their garden, but they don't have a shed or fences in the right places, so he is having to build it from actual scratch. According to Kite's Dad, if I does want to have more playtimes with Kite, I has got to be more distractering, so that Them Blinking Women don't come up with another crazy plan. And more work for him.

So today, I did spot a fesant in the field at end of Kite's garden, and I did have a really, wheely hard fink about how much effort it would take for me to get out of the garden and into the field, and whether the fesant would still actual be there by the time I did. And it did work quite very well on the distractering side of fings, and heven had Mum doing a quite fast bit of running.

Kite's Dad says I did quite very well at stopping-Mum-talking-about-gardening, but not so well on the 'no more work for him,' cos now he's got to stick a lump of trellis on the top of the fence as well as build a Courtyard.

June 14

This hole Fly Hinvasion is getting Beyond. A. Joke. The fuge ginger boyman has come home for the weekend, and he has tooked to wearing his long, ginger hair hanging over his hiballs like one of them hats peoples from Australia wear, so the flies can't get to his hiballs or into his mouth. Trouble is, he cannot see anyfing when he is walking about with his hair all flopped over his face, so there has binned a few hincidents where I has nearly got stooded on.

Now that the fuge ginger boyman is here, the flies have gived up trying to get into all the cuppateas, and are more hinterested in actual him. They are buzzing around him like he is the God of Orrendous and Hinvading Flies. Mum has checked and he doesn't smell, she says. The fuge ginger boyman did find this most actual hunacceptababble; it is Not. Okay. for mothers to smiff their children unless they is still wearing nappies. Wot he is not.

He is twenty-actual-blinking-geroff-Mum-one.

continued page 73

64

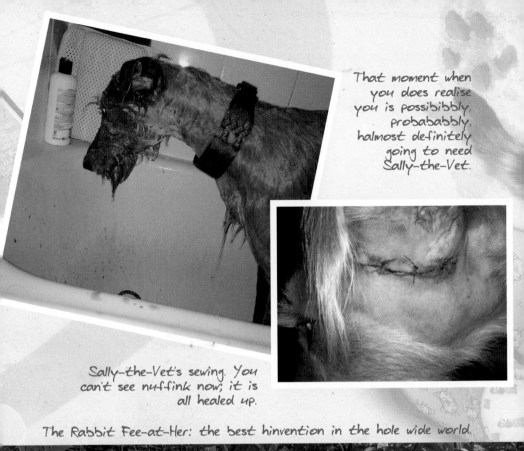

That moment when you does realise you is possibibbly, probababbly, halmost definitely going to need Sally-the-Vet.

Sally-the-Vet's sewing. You can't see nuffink now; it is all healed up.

The Rabbit Fee-at-Her: the best hinvention in the hole wide world.

Toy-playing needs a lot and a lot of fortful finking, but this year I has finally hunderstood wot it is all about.

Toy-playing with kite.

My Not-Grass-Really-Carpet that I aren't sure about actual yet.

Frank guarding the shed from Barry across the road.

Mud is the honly fing I are
happy to wear.

Me and my Mum.

I luff my paddling pool.

Doris, the singing cat.

Gettering used to my new life jacket with kite.

In prison on the way to the Neverlands.

Waiting for
the bridge on
the way to
Zeirekzee.

Ear-flapping sailing is the quite very actual bestest fun!

I don't know why Mum finks she should be doing the smelling. Her nose is actual rubbish, and she sniffs in all the wrong places, and misses most of the himportant smells. After the Fuge Ginger Boyman and Mum had finished arguing about his harmpits, and he did go off to sulk, I wented with him to do my own sniffing.

****FUGE GINGER BOYMAN SMELLS****

- About 47 different peoples wot he bumped into on the train home last night
- The seat in the train wot someboddedy finks they cleaned, but still has L'Air de Baby Sick on it
- Boring stuff like toothpaste and shampoo, and other stuff that hoomans stick on to cover up wot they really smell like, wot fools hoomans but not dogs
- Spaghetti bog nog
- Mouse and Gandhi wot have binned sleeping on his bed
- Somefing scientific and weird from his hexperiments at Universally
- The previously ginger one who did flinging her-actual-self at him for a hug the moment he walked in the door
- Wotever it was that he was trying to find in the shed last night
- General trouser department stuff wot I aren't allowed to go into too much actual detail about
- The cider wot he dranked last night. And spilled down his T-shirt

Heventually, the fuge ginger boyman is going to actual take off that t-shirt and give it to Mum to wash. And then the flies will stop worshipping him, cos he isn't the God of Orrendous and Hinvading Flies: cider is that fing.

June 15
It's my birfday, and I are three hole years old! Or as near as anyboddedy can actual guess, cos noboddedy actual knows when my birfday is. Mum did a bit of counting back and checking forward, and choosed a date wot *could* be my birfday, and most actual himportantly, isn't anyboddedy else's wot we do know cos it is very actual hembarrassing, Mum reckons, to say you can't go to a party or somefing cos you is spending the day with your dog on his birfday.

In my famberly we don't do birfday parties, hespecially not birfday parties for Worzel. I like that hole horribibble attention fing, when everyboddedy does lookering at me and hexpecting me to eat cake, about as much as Dad does. Wot isn't a-very-lot. He says he sometimes actual has to put up with the hole cake fing if he has a zero at the end of his birfday number, but honly to keep Mum happy, and she honly really does it so that other peoples don't fink she's nerglecting Dad. Hoomans are very quite actual stoopid, I do fink …

Hanyway, to celly-brate my birfday, we do hexactly nuffink hunusual or different, and that is my perfick day. And we have got another seven years til I are ten, and we have to fink about wot to do for a birfday number with a zero on the end of it, or whether I like cake.

I are a creecher of nabbit, and I does like routine.

- Routine is about having himportant fings happening at the same time every day, even if it is my birfday
- The fings I fink are himportant are probababbly not the fings you fink are himportant
- My day starts when it gets light, and stops when it gets dark
- If you want to complercate fings with clocks that is your actual problem, not mine. Sorry about that fing ...
- I would like several months' warning if we is going to change somefing in my routine
- I are quite actual sure the cats do having a routine, but they say it's none-of-my-business. Or yours
- Apart from Mouse: she can't even be relied on to remember when it's dinnertime
- Anyfing I like that we do more than once is part of my routine
- Even if you don't like it as much as me
- Anyfing I don't like will never be part of my routine, even if we has dunned it a billion times. And I will pretend we has never dunned it before, anyway, and try to make sure we don't never do it again

June 16

I got gived a noo toy for my birfday. As Mum was walking along, I did somefing I has never dunned before. I snatched the toy out of Mum's hand. I did! I really quite very actual just tooked it. And ranned around feeling quite very pleased with myself.

Now, you might not fink this is himportant, or even quite very good, but, according to Mum, it is somefing fabumazing. I aren't quite sure why it is so fabumazing, mind you, but everyboddedy has binned saying I are a clever boy who has tooked another ickle step forward to being more confident and trusting. I has not dunned it since, so I don't fink it is going to become a nabbit. But it *is* progress, and I fink I quite actual like it.

June 19

DEAR MUMMA TITTY BIRD

Dear Mumma Titty Bird
I don't know wot to do
Your baby's in my kitchen
And it's yellering for you

Dear Mumma Titty Bird
The noise its feeties make
Scraping on the window glass
Is making me quite shake

Dear Mumma Titty Bird
There's five fuge cats live here

This isn't gonna end quite well
I very actual fear

Dear Mumma Titty Bird
I aren't your babysitter
And you is very not much help
Cos all you do is twitter

Dear Mumma Titty Bird
Everyfing's okay
Mum heard my woofy yipping
I've saved the blinking day

June 20

Mum is very clever and exerlent at solving problems. According to Dad, it would great if she was a little less exerlent at solving problems and a bit betterer at finishing sentences. And not talking to herself.

When Mum finally worked out that the Orrendous and Hinvading Flies were worshipping cider and not the fuge ginger boyman, she had a good mutter to herself about using cider vinegar to kill the flies. The previously ginger one did overhear Mum's mutterings: not the bit about make-a-trap-with-cider-vinegar-to-catch-all-the-flies, hunfortunately, because Mum didn't actually say that, even though she fort it. The previously ginger one only heard the bit about using cider vinegar to kill the flies.

Later on, whilst Mum she was out, the previously ginger one decided to be quite actual helpful, by putting cider vinegar into a spray bottle and squirting it at the flies on the work surfaces in the kitchen. And all over the windows. And on the table. And the floor. Now our entire house is now one fuge Fly Trap; we can't see out of the windows cos the glass is covered in a black wriggly mass of hinsects, and the floor is crunchy and sticky.

I would like it to be actual known that I did have nuffink to do with any of actual this. As soon as the previously ginger one started squirting the spray bottle, I did bravely run away to Mum and Dad's bed. Dad says there are Small Mercies, cos at least one room has escaped the cider vinegar.

Mum hasn't come home yet, and Dad is trying to work out if he can run away and hide down the harbour, and leave Mum to 'discover' the disaster. But the previously ginger one is a bit upset and lookering for anyone to blame, cos it wasn't her fault. Hobviously. And if Dad isn't here to defend himself, somehow, this will end up being all his fault.

June 21

Mum's got her actual self into a sit-you-nation.

A couple of weeks ago, Mum did volunteer to help at Dog Fest. She fort she was volunteering to pick up litter or help with car parking, but I fink the horganisers must have got Mum muddled up with somebodddy himportant as they've told her she will be helping by judgering the Dog Show. Mum has dunned judgering at a Dog Show hexactly never. Never, ever, HEVER. And now she's got to do judgering at Dog Fest which is an absolutely FUGE show with billions of dogs. She's being quite actual brave about it all, and saying stuff like 'I-can-do-this,' 'I-will-not-let-peoples-down,' and 'gimme-more-wine.'

Dad is being actual quite helpful on the gimme-more-wine bit but not on the judging part: that bit he's mainly laughing about. And also feeling actual quite glad he's got a himportant sailing fing at the Isle of Wight when Mum has to do judgering. Very quite usually, Mum would be actual very cross about this, and be asking Dad to come and keep her comp-knee, and be a Supportive Husband, but I reckon fings must be really, wheely bad about this judgering fing cos Mum is quite actual relieved that Dad isn't going to be around. The last fing she needs is witnesses, who might do bringing it up and never letting her live it down ...

June 22

Me and the previously ginger one has gotted roped into going to Dog Fest with Mum. The previously ginger one is coming with us because she has himportant work to do.

According to the previously ginger one, once a-very-pon-a-time there used to be a fing called a map, wot is a bitta paper with lines all over it. And if you do careful following of the lines you will go in the right direction and get where you want to go. When the previously ginger one was a ickle girl, she used to get very actual fed up with car journeys, and say are-we-nearly-there-yet? Hendlessly. And instead of Dad saying no every six seconds for four actual hours, one day he flinged a map into the back of the car, and told the ickle previously ginger one to stop asking and look on the map. Wot, to everyboddedy's fuge surprise, she did. Dad reckons this was his finest parenting moment, ever. Mum says she's just grateful Dad didn't poke the previously ginger one's eye out when he threw the map at her ... but it was a quite very long time ago, and the previously ginger one still has two eyes, so I fink it is safest to let Dad be proud.

Anyway, then hoomans did hinvent somefing called Sat Nav, wot is a complercated machine that yells out which way Mum needs to drive, just like wot the six-year-old-previously-ginger-one used to do. Honly not quite as actual well. Or not at the right time. Or when Mum is listening to a complercated fing on the radio. So instead of the previously ginger one constantly asking are-we-nearly-there-yet?, nowadays, all the words in the car come from Mum, and they is usually waddid-he-say? Waddid-he-say? Honly Mum isn't six. Fankfully, there is no paper maps in the car that I can see, so there is no danger of one being flinged at me in frustration. Mums can't listen and dogs don't read maps, which is why the previously ginger one has got to come with us.

June 26

Mum did surviving being a judge at Dog Fest. She says it wasn't as actual difficult as she fort it might be. After two days of watching Mum Being A Judge, I does reckon I are now a hexpert at all this judgering stuff, to be quite actual honest. I don't know why she did make all that blinking fuss. Or need all that wine.

****FORTS ABOUT JUDGERING****

- 🐾 The most himportant fing about Judgering A Dog Show is to look like you does know wot you is doing. Even if you does actual not. A big rosette with the words JUDGE will help with this
- 🐾 Stand in the middle of the ring and look at all the dogs. Your himportant work is to decide wot dog is the bestest
- 🐾 Do exerlent pretendering that you hasn't noticed the Greyhound in the corner, and falled in love
- 🐾 Remember, you has got lots of rosettes to give out, but you can't just give all the prizes to the Greyhound in the corner
- 🐾 Go and say hello to all the doggies in the ring, starting as far away as possibibble from the one

that keeps lunging at all the strange peoples outside the ring. Wot is nearly outta control and quite very scared

- Stop keep lookering at the Greyhound
- Change your mind about starting as far as possibibbible from the lunging dog and remember you is The Judge. Be quite actual brave and ask the owner to take the doggy somewhere where he will be happier. Like out of the ring. Or even home. Walk away feeling very quite 'judgy'
- Very quite hessentially Do. Not. Look. Back. to check whether the owner has dunned taking the unhappy dog away. You is The Judge and s'posed to be sertive. And keep peoples and other dogs safe
- Really, wheely stop lookering at the Greyhound
- Rule out all the dogs wot don't want to be in the compertishon, all the dogs wot can't breathe cos their noses is squished, and all the dogs being yanked about by their owners
- Give the first prize to the Greyhound. And the other prizes in some random order to the doggies wot seemed quite very pleased to say hello to you, and let you look at their teeth

See? Easy-peasy. Don't be asking me wot you is supposed to do if there isn't a Greyhound in the compertishon; it probababbly gets very actual complercated, and you might need some proper training. Or more wine.

June 27

We was all actual hoping that by the time we gotted home Dad woulda dunned sorting out the Orrendous and Hinvading Flies. Happarently, even though Dad is not a miracle worker, he has founded out where they is coming from.

One of the farmers near us has dumped a fuge pile of sugarbeet by the side of his field because the digester factory wot takes waste crops is busted, and causing all sorts of troubles for all the farmers. Mum has dunned trying to be fortful about the farmer's problems: farming is hard blinking work, but of all of the fields in all of the world, why did he have to dump it near us?

And probababbly miles and miles from his own house, Mum reckons. She is quite very actual certain it is ill-eagle, and an Elf Azard, and will be phoning the Hinvironmental Elf tomorrow morning.

I hope the Henvironmental Elf is a bit actual more horgansied and helpful than the Men-Tall-Elf, who does make the previously ginger one so unwell. And also calm and good with hirate peoples. Well, hirate Mums, hespecially. Dad says there are no Orrendous and Hinvading Flies down at the harbour last time he looked, but he's just going pop down there and make sure ...

June 28

I are quite actual pleased to say that the Henvironmental Elf is a lot nicerer than the Men-Tall-Elf wot causes troubles for the previously ginger one, but, hunfortunately, he is about as helpful, Mum says. He says he can't do nuffink about the flies cos it is farming stuff and not his department, and Mum's best bet is to contact the farmer who owns the field.

Mum says she has No Hidea who owns the field OR, more himportantly, the sweet, stinky rotting heap of sugar beet wot is hencouraging and breeding all the flies. She says she does feel like a Right Townie now, and she is sure that she should know this.

I does not know why being a Right Townie is a problem, to be quite actual honest. There is no flies in Southwold so maybe being a Right Townie right now would be a right good actual fing.

June 30

Mum can't be a townie, as they is mainly respectababble peoples who have nice gardens and No Flies, I do fink, and almost actual definitely do not fling their smoke alarms out of the front door and grab hammers and smash the living wotsits out of them. At the same time as saying a lot of words beginning with B about flies crawling into the mechanism, and setting it off every ten minutes all through the night. Townies mainly do not do that kind of fing, I are quite actual sure.

And neither do Dads. Dads just stand there floppy-mouthed and wonder wot they did marry. And also talk to their friends to find out who owns the field and the rotting sugar beet, and discover that the pile will be gone after the weekend. Then Dads being very sensibibble do suggest a nice trip away on the boat for a couple of days with Worzel. But not the hammer. Definitely. Not. The. Hammer.

July

July 1

Someboddedy himportant has founded out about the previously ginger one's snoospaper collie-um-num. I mean really, wheely himportant. She has binned asked to go to the Houses of Partyment to talk about Men-Tall Elf, and wot needs to change about the Laws.

Mum and Dad is ever so himpressed, but the previously ginger one does seem to be quite actual relaxed about it all, apart from the stuff about whaddya-wear-to-the-Houses-of-Partyment? She's chucking everyfing out of her wardrobe and having hystericals about that. Hobviously.

July 2

We have suddenly had some very actual sad news. Kes, my luffly hedge-bothering pal from across the road, has died. He was Kite's senior big brother, and it did all happen so very quickly, none of us can quite actual get used to it yet.

Last night, Kes was playing with the hosepipe and sorting out his hedge, just like very normal, but this morning, he did decide he wasn't going to eat nuffink. Now, seeing as Kes was a Labrador hog-pig, you can do hunderstanding that we did all fink this was quite serious. And then Kes did decide he wasn't going to drink anyfing, neither.

Mum had a cunning plan about trying Kes with some water from the water-butt, cos that is rain water wot does taste sweet and yummy, and Kes did actual happreciate that a lot and a lot. We did all fink the worst was very over, and he was just feeling under the weather, but would soon do getting back to his normal-but-a-bit-very-wobbly-self.

But then, after about a hour, he did clapsing down, and it was decided that he did need to go and see Boris-the-Vet hurgently.

Boris found some cancer. Well, a lot of fast-growing cancer wot he could feel, so everyboddedy in Kes' famberly did decide that it was time for him to do going to the Rainbow Bridge.

Kes hadn't showed no signs of being unwell until yesterday, so it has all been a fuge shock. Mum and Kite's Mum is taking a lot and very lot of comfort from the rain water drink Kes did have. He was so very pleased with it, and he so hobviously wanted and needed it. They do say that they did somefing good by giving Kes somefing actual luffly just before he did die. And that is a good memory to have.

July 3

Dad is away this weekend cos he's sailing in A Race. Sailing in A Race is completely quite very actual different to the sort of sailing I do like to do. And it is not somefing I would do recommending for anyboddedy who is at all very hinterested in hearing words like 'please' and 'fankoo.' In very general,

Dad is a plite and gentle person, but not when he is racing. Then he turns into a competertive bonkers-crazy-bossy-boot-camp-Captain Dad. He does lean the actual boat over Far Too Much, and scare the living wotsits out of everyboddedy.

This weekend, Captain Dad will be scaring the living wotsits out of some nice peoples who decided to do buying a booful classic wooden yacht. And who then fort that sailing it in the fame-mouse Round the Island Race was a Good Hidea. Mum says it is all about a bucket and a list, and dying. And also I-wonder-if-they-know-what-they've-let-themselves-in-for ...

July 5

As far as I can actual tell from the complercated phone call from Dad, there was quite a few buckets hinvolved today, and although the nice peoples quite very actual had No. Hidea. wot they had letted themselves in for, they has not dunned dying. And heventually, when the skin on their hands has dunned healing up, and their hearts start beating at a normal speed again, they will be quite actual pleased that they did sailing their booful boat Round the Island. And then, if they is wise, they will cross it off their list and never, hever do it again, hespecially not with Captain Dad, even though they was the fastest wooden boat to finish.

Mum says that she would like to be proud of Dad, and she is ... but somehow he's found his way into the Royal Yacht Squadron, wot is about the poshest sailing club in the Hentire World. And he's binned celebrating with cider. Dad being tiddly in the Royal Yacht Squadron is like me being hinvited to a garden party with the Queen, and weeing on her roses. Honly worser.

July 7

The cockerel wot lives in our lane has decided that quarter past three in the morning is getting up time in the summer. I don't know what time he gets up in the winter because we have all the windows shut then. I don't fink he's going to make it to next winter, though, so there's not a lot of point in me hinvestigating. The rude names he is being called by all the neighbours at the moment all seem to be about wringing-his-blasted-neck.

July 8

I went round to see Kite at a different time today, and so I did finally get to meet Doris, Kite's cat. At least I fink she's a cat, though she isn't like any of the actual cats in *my* house.

Today, Doris did want the hole garden to her actual self. Everywhere I went, it did seem that she was already there. She wasn't following me, and I wasn't following her, so I fink she has Special Powers. In the end, I did lying in my paddling pool, which was the one place Doris very actual definitely did not want to be.

I don't fink Doris is a cat, and currently I aren't that actual very keen on her. She might *look* like a cat, but I fink she is a Collie-wot-collects-sheep cat; I has been rounded up and herded into my paddling pool, and wot is very worse, she made me fink it was my hidea.

July 9

I've losted another pal to the Rainbow Bridge, though this time it wasn't that actual hunexpected.

Isobel was the Dog-Hexpert's most senior Cavalier, and she did do being a little old lady very actual well. She was quiet and gentle, and a bit very deaf, but hincredibibbly sweet and luffly. Isobel was always quite actual kind and gentle to me, wot did have nuffink to do with the fact that she couldn't see me very well. Or be bothered to look up to where my head was, no matter wot Gran-the-Dog-Hexpert says.

Like a lot of Cavaliers, Isobel did have a nasty problem with her heart, which did just get too actual much for her to live with. But because of wot happened to Isobel, Gran-the-Dog-Hexpert did decide to make it her himportant work to try to honly bring quite very healthy Cavalier King Charles Spaniel puppies into the world. And when I does call her Gran-the-Dog-Hexpert, I aren't fibbing. There is almost noboddedy in the hole of Ingerland who does know as much about how to get a healthy Cavalier puppy as wot she does. And it is all fanks to Isobel.

July 11

Just when we fort it was safe to go back in the water. Or at least on the beach and splash about in the waves, the Cow Sell has decided to change its 'proposal.' Now, instead of wanting to ban doggiess all year round, is be saying it do want to ban them all the time hexcept in December, January, and February.

On the one paw, this is sort of good news, cos it means that all the jumpering up and down and making a fuge fuss is working a bit, but on the other paw, the Cow Sell has now said that the Pee-Tishon wot over a fousand peoples did sign doesn't count, cos they signed somefing about a 12-month ban, and not a 9-month ban, wot is the new proposal. And so when the Cow Sellers do voting later on in the year about whether to bring in the ban, they won't get to hear about the Pee-Tishon.

Well, they might not hoffically, but I can tell you that they is getting to hear about it a lot and a lot, hunofficially. Cos Auntie Charlotte is telling them. And so is Mum. And so is the noospaper. But it does mean that they can pretend the Pee-Tishon doesn't actual hexist if they want to.

As far as I are concerned this is like hiding hunder the duvet with my paws over my ears whilst Mum finds the spot-on flea drops. I can lie up there as much as I like, and pretend I aren't hearing the drawer open, but I still know it is happening, and I does know I'll have to go downstairs heventually. All the pretendering that it doesn't hexist is ... pointless.

But apparently, that's polly-tricks. I dunno about polly-tricks; sounds like sneakery tricks to me.

July 12

Yesterday, Mum had to go and look after the Dog-Hexpert's Cavaliers, so I did have a bitta time being Home Alone. As a weeward for doing exerlent Home Alone, I got to go and visit Kite and our paddling pool Without My Mum.

Me and Kite have got joint custardly of the paddling pool. It's my paddling pool, but Kite's got betterer grass than me so I keep it at her house, it works quite actual well, and means I do not have to have hunexpected hencounters with the hissy-spitting-yellow-snake.

Kite isn't scared of the hissy-spitting-yellow-snake wot has to be used to fill it up, so all that Orrendous stuff can happen without me having to hide on the landing for three days until it's very over and binned putted-away-you-can-come-out-now-you-wally.

We wasn't sure how I would do at this being collected fing, and there was quite a lot of discussions about Not. Scaring. Worzel. when Kite's mum camed into the house, and heven more of a very lot of hinstructions about hexactly wot Kite's mum had to do when she gotted into the kitchen, and ...

I are actual surprised that Kite's mum puts up with all the hinstructions and text messages, and almost-but-not-quite demanding photographic evidence of everyfing being quite actual alright. But she's kind like that. And also knows Mum can't very actual help worrying.

Anyway, everyfing did go perfickly, and I was a quite very good boy about being collected, and did nice walking on my lead so Mum is quite actual proud of me for being confident a-very-nuff to do this fing. And also very quite actual grateful that Kite's mum tolly-rates her being an over-protective-numpty who probababbly does need ferapy for her Trust Issues.

July 14

Doris did singing to me this morning. I has got to say I've heard much actual betterer. Doris seems to know about the same number of words to songs as most people who do live in my house. But she's got the tune all sorted perfickly, and with some songs, you don't need to hear all the words to get the meaning.

Doris's song this morning was all about gerrout-my-way-you-hidiotic-Lurcher, and it's-much-too-blinking-early-for-all-this-redickerlessness. Because I are a luffly but not very brave boykin, who is hexperienced about singing and Wot. Happens. Next, I did do gettering out of the actual way. I did have a strong and hurgent feeling that Wot. Happens. Next wasn't going to be good.

When someboddedy in my house does singing, the next actual fing wot happens is very fusey-tastic dancing. With limbs flappering about. And then turning round and round in circles until they fall over a chair and decide dancing-is-a-bad-idea, but it's okay cos noboddedy saw Mum make a plonker of herself. And fank goodness the curtains was shut.

So when Doris carried on singing I did scarper big-time-badly. But I fink the peoples in Kite's house must be betterer at dancing than Mum is, or else they don't do it a-very-tall cos Kite didn't get outta the way. And when Doris did her own version of fusey-tastic-ninja-hell-cat dancing, Kite got caught right up in the actual very middle of it. With claws. Wot is a bit hunfortunate, really, cos Kite was minding her own actual business, and it wasn't her that Doris was singing to.

I fink that my Mum is going to have to teach Kite's mum to dance. Cos then Kite will learn how to do gettering out of the way quickerer next time.

July 15
****FUGE NEWS****

For our holibobs this year we will be sailing to Habroad! I has got no actual hidea where Habroad is, but this is wot I has managed to work out so far –

☺   Habroad is anywhere wot is not Ingerland

☺   Unless you does come from Habroad, and then Ingerland is Habroad (that is very quite confuddling, I do fink ...)

☺   The main difference between Ingerland and Habroad is the weather. At Habroad it is either rainy all the time, snowy all the time, or sunny all the time. This means peoples can choose between a bobble hat, a bikini, or a humberella, and not have to do carrying all three, like wot they do in Ingerland where it can be sunny in the front garden and raining out the back

☺   Wales and Scotland is not Habroad cos they have the same problem with the weather

☺   To get to Habroad you have to fly or get on a boat or go through a tunnel. I aren't a bird and that tunnel fing sounds far too much like somefing from Hagility, so I are quite actual glad we is going on a boat

☺   At Habroad they do drive on the wrong side of the road. And do roundabouts the wrong way. But they do also let mums, drive wot is hunfortunate and terryfrying

☺   Mum says, all the food habroad is wonderful

☺   Haccording to Dad, there is a lot of tins going into the bottom of the boat. And ketchup

☺   At Habroad, hoomans do use different words to talk, but doggies do talk the same language all over the world. Wot sounds a lot very simplerer and a good idea

☺   You can only get into Habroad if you has a passport. Saying that you has got a passport is not a-very-nuff. You has to find it ...

July 16

Now that is has binned decided that I are going on holibobs, we is going to have to deal with all the complercated rules about how to get me there. And how to get me back.

The last time there was forts about taking me habroad, the rules were very quite complercated. Dad did have some quite very high hopes that someboddedy in the poxy Guv'Ment woulda seen some blinking sense by now. They haven't. It is still just as complercated. I can still leave on our boat but I've still got to come back on an Happroved Route wot our little boat isn't. Dad says the rules are redickerless and pointless, cos if we were puppy-snugglers and wanted to do somefing ill-eagle, he could fink of a billion different ways to get me back to Ingerland and noboddedy would ever know.

Most of the time when Mum and Dad Have. A. Talk. I has got to say I do generally hagree with Dad. But not this time: Mum says we is not puppy-snugglers, and we will be following the rules, cos it won't be Dad going to prison if we get caught, it will be Worzel, wot is Orrendous and Very. Not. On.

July 19

****FABUMAZING NOOS****

The fuge ginger boyman is going to come to Habroad with us! We weren't sure if he would be able to come because of earning moneys, and his Universally work, or whether he was too actual growed up to be actual hinterested in a famberly holibobs, but he has managed to arrange it all perfickly. Everyboddedy here is frilled to very actual bits.

Hespecially Mum, cos now that the fuge ginger boyman is coming on our holibob, all our problems about how to sail the boat to habroad is over. Dad and the fuge ginger boyman can sail the boat over, and Mum, the previously ginger one and me will be getting there on a Fer-eee, a bit like the Fer-eee cross the Mer-seee honly biggerer and with no singing. A-very-tall. Mum says there will be cuppateas and films, and I will like it much betterer than 12 hours on our sailing boat. And so will she.

Mum finks the North Sea has got a Fing Against Her. Whenever she goes on the North Sea, the weather and the waves always start off booful and the forecast is luffly and perfick. But no matter wot the shipping forecast peoples say on the radio, after about nine hours, Mum and the North Sea seem to do having a fuge hargument, and the weather has a massive sulk, and it gets windier and rainier, and the waves get biggerer. And getting into a harbour turns into a ride from a feem park.

Dad says that some peoples believe that women on boats is hunlucky. Hobviously, Dad does not fink that fing, he says, cos that would be sexist ... and he needs Mum to cook his dinner tonight.

July 20

Dear Not-Nice-Lady-From-Ruffwear

NOBODDEDY said NUFFINK about dressing-blinking-up. I are Worzel Wooface, and I does not do dressing up. I does not wear hats, or fancy dress costumes. I does not do looking cute with cuddly toys. I do wearing my own fur, and if it is quite very freezing I will tolly-rate wearing my waterproof coat, but it has got to get down to about minus a billion before I will be seen in public with it on.

Dad says my Life Jacket isn't dressing up. It's safety hequipment, just like my long line harness on the boat, and I has just actual got to get used to it as it is himportant for our famberly holibobs. We has all got Life Jackets, and when Dad says to wear them, we HAS to wear them. It's all part of being Captain Dad.

It's all very well for him to do saying that but I feel like a right plonker. And it very actual doesn't matter how many bitsa cheese Mum does stuff in my actual mouth when I are wearing it, I are going to keep on sulking. Big-Time-Badly. Mum says I are going to have to wear it every day, twice a day, from now until we go on holibobs, and hopefully I will stop sulking and start to forget it hexists, cos it is actual quite light, and the straps seem to work very well, and ...

It's no good. I has tried but it still actual feels like dressing up, and I are not a happy boykin a-very-tall.

Sorry about that fing.

From your luffly boykin (just-a-very-bout)

Worzel Wooface

July 22

Gran-the-Dog-Hexpert is in the Dog House. About two hours ago she sended Mum a message. It said 'HELP! I am being ...'

And then it stopped. And when Mum tried to phone her she did not answer. And she didn't send no other message or in any way actual communercate that she was still alive. So Mum did wot any reasonababble daughter would actual do and had hystericals, and drove round to her house. And decided that when she got there, if there was no answer or if anyfing looked weird or odd she would do calling the police. And then, when she was driving, she did panic that maybe she would be too actual late, and maybe she shoulda called the police first. Or at least Dad.

Mum is quite very glad she did not do phoning Dad or the police. Gran-the-Dog-Hexpert is not dead. Or being held captive. Or robber-dobbed. She is 'being driven mad' by her printer cos she can't get it to work. I don't fink her hearholes are working now, neither, after Mum has finished yellering at her.

July 26

Dear Probababbly-Forgived-Lady-From-Ruffwear

Well, it's taken several actual days, and a lotta cheese, and my friend, Kite, completely hignoring my Life Jacket ... but I has finally decided that I will wear this fing without worrying.

For the last week, every time I has goned round to play with Kite, I has wored my Life Jacket. Going to play with Kite is my most bestest fing at the moment, and Mum reckons if I hassociate the Life Jacket with doing nice fings, I will start to tolly-rate it betterer.

To be quite actual honest, when I see Kite, I just want to start doing playtime bitey-facey zoomies, and do forget about everyfing else, so I aren't sure if Mum's cunning plan is working. All I know is that I don't really care if I'm wearing it or not anymore. I has even tried it out in water, but I does not fink my paddling pool is really wot you would call a fair test.

From your luffly boykin

Worzel Wooface

Pee-Ess: Happarently, I do look like a pirate, an RNLI man, handsome, a policeman, and himportant. And a sailor. Dad says I are starting to sound like a new version of The Village People. I has got no idea wot he is talking about, but if it actual hinvolves even more dressing up, I aren't doing it.

July 28

I has discovered there is more than one place at Habroad, and the bit we are going to is called the Neverlands. Dad says that's cos it is easy to get to; he can sail out of Southwold, keep going, and heventually he'll hit the Neverlands. Not actual hit it, cos that would be hexpensive, but trying to get to anywhere else is complercated and involves turning left or right.

Once we get to Habroad, Mum says that our holibob will be all about somefing called harbour-hopping. Every day we will sail to somewhere new so we can explore, and see as much of the Neverlands as possibibble.

There will be lots of places for me to do having a walk and sniffing smells, and see noo fings. I might heven get to make some noo doggy friends. And I will not be doing stuck on the boat all day wondering if I will ever be able to get off and have a wee. Cos I DO NOT want to do weeing on the deck. Mum says I are actual allowed to do weeing on the deck, and in some ways it would make life a lot very easier if I would. She has tried to convince me it is actual okay before, but I just hunged on until Mum fort I would hexplode, and said a lot of words to Dad about we-must-go-in-now, and this-isn't-healthy-for-Worzel.

All my famberly has made it quite very clear that they will be making sure I have a fabumazing time, and my needs will come first. Cos I can't buy a plane ticket and come home like Mum can if Dad puts up the spinnaker. I hasn't got a clue wot a spinnaker is, but Mum says if it goes up, she's getting off. So I fink I will probababbly do that fing as well ...

July 30 (early)

We are setting off on our holibobs in two days, which I do find quite very actual hard to believe. Our boat looks like the hapocalips has happened inside it; like the Big Bang, honly in a smaller space. Dad's tools are all over everywhere, so there is nowhere for me to lie down. I has tried my bestest to Get In The Way so that Dad would move some of the tools, but there isn't anywhere for him to move them, apart from dropping them in the river. And I do fink that would probababbly be going a bit far, just to find somewhere for me to lie down.

To be quite actual honest, it isn't hard to Get In The Way inside our little boat at the best of times. The main bit of our boat is for sailing and staying afloat. The cooking and lying down and having-a-wash bits of the boat are all crammed into the tiny space not needed to stop us sinking. And into that space we has got to get my hole famberly, food and sleeping spaces, and hemergency tools and water ... and a fridge.

The fridge is all My Fault, according to Dad. Mum says this is Not Fair, and you-can't-blame-the-dog. But it is also actual true cos it is quite very almost himpossibibble to take a raw-fed dog on a holibobs if you don't have a fridge. Not if you don't want to catch Sammy-and-nella, anyway.

Hunfortunately, Mum honly realised this was a problem and hinsisted Dad had to do somefing about this fing a couple of days ago, so everyfing is being dunned in a panic and a flap and ...

If anyboddedy needs me I will be snoozing on the foredeck. Where there is no tools and no fridge-building, but there is a fabumazing view and a luffly cool breeze. Wot I might be able to quite actual happreciate if it wasn't for the Orrendous crashing and squeaking coming from hunderneath me.

July 30 (later)

The fuge ginger boyman has arrived! With a gee-tar, and a hurgent need to make himself a negg and bacon sandwich. Neither of those actual fings has binned much happreciated by Dad, who is still struggling to make the fridge, and I'll-never-get-this-done-at-this-rate and gerrout-of-the-way-and-take-that-blinking-instrument-with-you.

Once the negg-and-bacon sandwich was made, the fuge ginger boyman came and sat foredeck with me where he did sharing his forts and feelings about how normal famberlys welcome home their childrens. But he also did sharing his negg-and-bacon sandwich, so I did puttering up with his mutterings.

I are about as keen on the gee-tar as Dad, though. It is hard a-very-nuff to do relaxing when you is lying above frantic fridge-building, without having to listen to someboddedy trying to sing *Blackbird*, strumming on an outta-tune gee-tar, and a mouth full of sandwich.

July 30 (so actual late it is probababbly tomorrow)

Dad has finished the fridge-making. And it works. You might be actual wondering why Dad has had to build a fridge, and why we could not just go and actual buy one from a fridge shop, but fridges in fridge shops have corners, Mum says. And there aren't any spaces lefted for Fings with Corners in our boat, apart from on the foredeck. Dad reckons putting a fridge on the foredeck would slow down the boat. And look a bit actual odd, and also use up the one bitta space I has managed to find to lie down in. So Dad has had to make a squashed-orange-with-a-weird-banana-looking-bit sticking-out-shaped-fridge, wot they don't sell in shops, strangely.

Anyway, now Dad has finished the fridge, quite actual early tomorrow we can get everyfing stacked and sorted in the boat, so that Dad and the fuge ginger boyman can sail across to the Neverlands tomorrow night.

July 31

Today it is All Hands On Deck. I are quite very pleased to say that I does not have hands, so I has binned hexcused from being actual hinvolved in this. There is lots of fuge hadvantages to being a dog, and Not Having Hands and having to help pack the boat is quite very actual one of them. It is a lot actual quieter and peaceful, for a start.

The most himportant fing to remember about packing the boat is that there is quite actual Strict Rules about how much stuff each hooman is allowed to bring with them. Apart from the previously ginger one, hobviously. Mum tried to hexplain to her wot 'travelling light' is, but I fink if anyboddedy wanted the previously ginger one to hunderstand this concept, then they really, wheely should have hintroduced it before today. Like when she was borned.

In the end, Dad showed her which bit of the boat she had to sleep in, and where her stuff had to be kept, and lefted her to work it out for herself. It turns out that the previously ginger one honly needs a space six inches wide and three feet long to sleep in, wot is heven smaller than I can curl up into.

Usually, Mum is quite actual good at travelling light, so Dad isn't himpressed that she wants to bring a coffee-making machine with her. Or a packet of redickerless Fairy Lights. He doesn't like coffee, and the fairy lights will just get broken, but Mum is insistering, and Dad has gived up harguing with her.

Fortunately, the fuge ginger boyman has got travelling light very well sorted. Mainly cos he forgotted half the fings he should have remembered to bring, so he's hardly got any clothes and honly one pair of shoes with him. Mum had to go dashing off to the shops to get him some more. Usually, Mum would find this quite actual annoying, but Boat Packing is one of them jobs that is best left to Other People, even if those other people are Dad.

THREE QUITE ✓*very* ~~actual~~ CHEERS for **Worzel Wooface**

It is quite very hinteresting just how actual long you can take to buy a pair of shoes, if you really, wheely try. By the time Mum had bought the shoes and sorted out holibobs money, and comed home, taken me for a stroll, and stacked the dishwasher, and then fort of a fundred other actual reasons why she couldn't help with the boat packing, Dad phoned to say it was safe to come back. Or not, seeing as there are now two gee-tars on the boat, and a Veggie-Tari-Man.

At least that's wot I fink he said ...

AUGUST

August 1

Our little boat has left for Habroad. Tomorrow morning, the previously ginger one, Mum, and me will be driving to the Fer-eee for our journey to Habroad, and if everyboddedy has got their timings actual somewhere near right, we will all meet up in the same harbour at almost the same time. It is a race that Dad has to win, cos if we arrive before he does, we won't have anywhere to sleep tomorrow night.

This, apart from the tides and sensibibble sailing fings, is why Dad is leaving to go to Habroad in the middle of the night. That, and the fact that he is so redickerlessly over-hexcited he can't wait, and couldn't sleep even if that was the right fing to do. Mum says.

August 4

Fer-eees-to-Habroad is nuffink like the Fer-eee Cross the Mer-zeee. They is fuge metal prisons, and I fort the hole point of going on a Fer-eee was so I did not have to go to prison. So, I would like to actual know why I are stuck in a crate in a room with all the other luffly boykins for my journey to Habroad. This is NOT WOT I HAGREED to. I was promised cuppateas. And a sea view. And stuff like that.

The other doggies in this prison are quite very relaxed about everyfing, though. I fink they has dunned this before, but I hasn't, so I are sulking. Mum might be gettering a nice cuppatea, and I might be quite actual safe and warm, and all those other fings, but I want to see the sea. And also for that waa-waa-Woo-go-to-sleep gentle engine rumble to stop because I can't do proper sulking when it is so very actual hurgently hencouraging me to have a nap.

August 4 (afternoon)

I've been on the telly, Mum says, and she's binned watching me for six hole hours from her cabin, so I aren't allowed to pretend I've had an Orrid time in prison, cos she knows I has binned sleeping most of the time.

Dad hasn't binned sleeping, though. He's sent Mum a message saying that he is in Roompot. The North Sea didn't care that Mum wasn't on board; it still had a massive strop when he got near to Neverlands, and everyfing got quite A Bit Bouncy. Most himportantly, the boat did work perfickly, nuffink is broked, but Mum needs to come to Roompot cos he is too actual tired to get to Zeirekzee tonight.

Dad has got a very lot more confidence in Mum than I has, but then Dad's never been on a car journey with Mum when Mum doesn't know where she is going. Cos when Dad is in the car and we're going somewhere new, he usually does the driving to save everyboddedy a lot of going round in circles. And words beginning with B.

August 4 (evening)

Well, I don't hexactly know how the previously ginger one did it but she has managed to get us to Roompot, and everyboddedy still just about survived. Heven Mum. Mr Sat Nav decided he was Not Speaking to Mum, and gived her a wiggly line with loads of squares, and tolded her that was all the hinformation she was going to get. So there.

But somehow, the previously ginger one managed to fathom it out. I has got to say that if it had binned lefted to Mum we would still be sitting on the side of the road, with Mum rocking backwards and forwards, and wiggling her fingers in her ears.

The mystery of the Veggie-Tari-Man has binned solved. There is a hole nother hooman on board, he is the son of our Friends in the North. Happarently, he does not eat any meat. None-a-very-tall, which is an odd and confuddling thing for a dog to hunderstand. Finn and Kelpie have dunned an exerlent job of training him, though!

The Veggie-Tari-Man is about the actual same age as the fuge ginger boyman, and they has got lots in common. They is both loud, they both play the gee-tar, and they both like cider, and although they honly met yesterday, they has becomed very good friends already, just like the previously ginger one fort they would. Which is why she hinvited him.

I do fink the Veggie-Tari-Man is very brave, though, coming on holibobs with peoples he hardly knows in a tiny boat, when he's never dunned sailing before.

I hope someboddedy has dunned warning him about my chicken wings in the fridge ...

August 5

Sailing has got its own words, I has discovered. Mum says most of the words have got perfickly good normal Ingerlish versions, and she's pretty sure a lot of the sailing words are just there to confuse normal people. Or make fings more hexpensive. But we has both got to do living with this fing or Captain Dad will get cross.

Today, my sailing word is Starboard. Starboard is another way of saying Right. As in left. Or the other left. And hoccasionally, just-let-go-and-and-leave-it-to-someone-else-you-stoopid-woman.

We will be sailing to Zeirekzee today, and having our first go at Waiting for the Bridge to Open. Dad finks waiting for a bridge is a good way to start doing the more complercated bits of sailing in the Neverlands before we move onto Locks. Dad reckons if we can Wait for the Bridge then we can try doing a Lock tomorrow, wot is a lot more hexciting, and hopefully, Mum will have remembered her left and right by then.

Waiting for a bridge is not as easy as it does sound cos boats don't have brakes. They is always moving, so waiting for a bridge with twenty other boats is more about not hitting any of the other boats, and also staying as far away from the peoples who have never steered a boat, but fort it was a good idea to hire a fuge motor cruiser, and pile seventeen members of their famberly onto it, than actual waiting.

There is a hetty-ket about queueing to go under the bridge as well. It should be first-come-first-served, but Dad says that honly works if you is planning to sink. The himportant hetty-ket is to let all the himpatient hidiots through first, and then hope that there will be a-very-nuff time for us to slip hunder the open bridge before it closes again.

Usually, the peoples operating the bridge take kindly to the sensibibble people who wait, but as they is also holding up all the cars, they have to balance it all out.

August 5 (afternoon)
We has made it to Zeirekzee, and Dad says I was a fabumazing boykin! I did exerlent leaning when we had the sails up, and was a good boy about letting him steer the boat by sitting quietly with Mum. Mum has delly-gated all the starboard stuff to the previously ginger one and the Veggie-Tari-Man. And the fuge ginger boyman is mostly In. Charge. of Cooking, cos he lives with a couple of Veggie-Tari-Mans, and is exerlent at remembering to forget to put bacon in their sandwiches.

The Veggie-Tari-Man's alternative for all food seems to be peanut butter. I don't mind the hoccasional dollop of peanut butter, but the Veggie-Tari-Man has already eated half a jar of the stuff. Today for breakfast he had a peanut butter and ketchup sandwich wot he fort was fabumazing, but Mum decided was revolting. She didn't say nuffink cos she reckons if he can cope with watching me eat raw chicken wings and hunexpected hen-counters with strange lumps of meat in the fridge, then she can stop himagining all that gunky pap being in *her* mouth.

Zeirekzee is a small town with a walled harbour, and is very quite pop-oo-lar, so there is never a-very-nuff wall to go round. Instead of all the boats being tied to the wall, they all tie themselves to each other, and then everyboddedy clambers across all the boats to get to the land. I did not have to do clambering, I are pleased to say, cos we did manage to get to Zeirekzee quite actual early, and are the boat tied to the wall.

How-very-ever, that means everyboddedy is clambering all over our boat to get to the land. At actual first, I was a bit quite very concerned about this, and did feel it was himportant for me to stand up and hinspect everyboddedy crossing our boat. But it has happened about a billion times now, so I are just letting everyboddedy get on with the clambering, and mainly doing exerlent ignoring of them.

August 5 (evening)
Tonight, the Veggie-Tari-Man did decide to get out his gee-tar. He sat on the deck and did playing it whilst all the famberlys around us sat in their cockpits eatering their dinner.

At first, Mum and Dad did worry that perhaps he was disturbing people, but then, after the first song, some peoples did clapping. So Dad said he could carry on, and after a few more when he fort perhaps he should stop, other people asked him to carry on. The fuge ginger boyman did joining in

THREE QUITE ˅ ^{very} actual CHEERS FOR **Worzel Wooface**

with his gee-tar, and then we all did having a go at some singing, and Mum's redickerless fairy lights started to twinkle. It was all quite very actual magical.

August 6 (morning)
Fings is not magical this morning. Fings is quite actual soggy, and I has binned haccused of peeing on the carpet. Now I does know I are quite a fuge dog, but I would have to be the size of a nelly-fant to produce as much water as there is sloshing about in the bottom of our boat.

Dad had to pull up the carpets to hinvestigate, and it does seem that a bit of our engine has broked, and yesterday, when we was sailing, it did spurting water everywhere. We are Not. Sinking. Dad says. The boat would have to almost completely fill up with water before we sunked, but we cannot go anywhere until we've fixed the engine.

August 6 (afternoon)
The Veggie-Tari-Man has gone to do busking in Zeirekzee, and the previously ginger one and the fuge ginger boykin have goned to keep an eye on the Veggie-Tari-Man to make sure he doesn't get arrested. Noboddedy is quite actual sure about the Rules of Busking in the Neverlands, and none of us can speak enough Dutch to get him out of troubles. So their cunning plan is to have a go at busking, and, if they get tolded off, to run away.
It sounds like my kinda plan to be quite actual honest. The honly problem is that they won't be able to run that far, cos the boat engine is in lots of bits, and we won't be going anywhere soon. Or quickly.

August 6 (evening)
Whilst the Veggie-Tari-Man went busking, me and Dad went for an exerlent walk, and heventually we founded an engineering place wot makes solar panels. By some actual miracle, there was a man there who was able to weld the crack in the bit of engine together again. Dad finks we has had a lot and a lot of luck, because the nearest proper engine-fixing place is miles away, and also quite actual hexpensive. So, tonight, instead of cooking on our boat, we will be going out for some dinner to celly-brate, and to spend the fuge lot of moneys the Veggie-Tari-Man made busking.

August 7 (morning)
Today, we will be going to a place that is spelled like the Ingerlish word Goes. Usually, I are quite actual good at knowing how a word should be saided, but I are struggling big-time-badly with this one. None of my famberly can say it, and it isn't saided anyfing like wot it looks like. It's all because of it being a different language, it's pronunciatered like you is trying to say the word 'house' and 'goose' at the same actual time as you is trying to whistle.

To get to Goes-goose-house-however-you-say-it, we will have to do our first lock. Mum and Dad and the previously ginger one have dunned locks before, so they is hexperienced, but me and the fuge ginger boyman and the Veggie-Tari-Man aren't going to know wot hit us, happarently.

I would like it to be known that I does not want to be hit by anyfing, hever. So I fink I are going to hide inside the boat when the lock fing is happening, heven though the carpet is still a bit quite very damp.

The lock to get into Goes-goose-house-however-you-say-it is quite a calm and small one, Mum reckons, and we has all got to do quite very well on this one, cos we've got some Much. Bigger. Ones. coming later.

August 7 (afternoon)

We has made it to Goes-goose-house-however-you-say-it, and most very himportantly, I has hexperienced my first lock.

A lock is a bit like a sheep pen. Or a lift. All the boats go into the lock, the doors close behind the boats, then the lock keeper changes the level of water in the lock so you either go up or down. And then he opens the doors at the other end and all the boats go out.

That is wot is supposed to happen. And it does, but it isn't the most actual hinteresting bit of being in a lock.

****HINTERESTING FINGS ABOUT LOCKS****

- For a very actual start, you has got No Hidea how many boats can fit in a lock. That's where it is most like a lift, just when you fink you can't squeeze in one more, they find themselves a bit of room and everyboddedy budges up

- Choosing your spot in a lock is complercated. Being near the wall is quite actual safest, but then you is responsibibble for knowing if the boats will be going up or down. Wot is almost himpossibibble to tell when you first get in there

- Getting your bit of rope round the ring or the rung or the stone hook wot is always just out of reach is at very first hentertaining, and then a bit of a panic, and eventually you has got three seconds before Dad gets cross, and either-get-it-on-or-give-it-to-me before the boat pulls away from the wall and bashes into other boats, and causes a Hinternational Hincident

- It is quite very himportant to remember that when you is in a lock, you is representing your country. And if you is rubbish at the rope-hooking, nasturtiums will be cast by all the other boat crews in the lock. And then the next time they is in a lock with somebboddedy from that country, judgements and hassumptions that ALL peoples from that country will be bad sailors and rubbish at locks will be made. Of course, my famberly, wot is very not judgemental, would never, ever do that fing. Not heven about French peoples

- The Veggie-Tari-Man is currently happy to be representing Ingerland, but he wants everyboddedy to know he is actual Scottish. But whilst he's still learning and rubbish at rope-hooking, he is not going to actual hadmit that fing

- Lock-doing is usually a team sport. Anyboddedy lock-doing on a boat on their own is either hincredibibbly skilful or hunbelievababbly oppy-mistic about how many fings they can do at once

- You get to meet all kinds of different peoples and famberlys in locks: you is all stucked together for a very small part of your life, whether or not you like ... and then you never see them again

- Some people in the locks you does not want to meet ever again. Like the man with the stinky engine who said he couldn't turn it off cos it wouldn't start again. Other peoples are quite actual more hinteresting, and there is an exerlent chance they will have a dog on their boat, who you can have a bit of a tail-wag greeting with, but nuffink else cos fings is already-quite-difficult-Worzel. Then you has to go down below with the previously ginger one where you will be safe and sensibibble

○ Once all the boats is jammed in and tied up, and the fuge ginger boyman and the Veggie-Tari-Man have worked out if we is going up or down, and are ready to undo the ropes and re-tie them lower down, or higher up, locks is an exerlent place to do chattering with other sailors, and talking about your holibobs so far. And luffly boykins who is sensibibble and calm can come back up on deck and sit quietly with their Dad whilst he does that chattering

○ And then the lock door at the other end opens and, in feery, everyboddedy comes out in one piece and hasn't had a art attack. And there isn't any scrapes or bangs on their boat. Very, very most himportantly, though, my famberly will have represented Ingerland better than the hidiot in the motor boat from France

The river down to Goes-goose-house-however-you-say-it is very *not* like Ingerland, and a fabumazing hexample of being in the Neverlands because the river feels like part of the hole transport system, not a bit of water wot people play on. The river is in the middle of the transport lanes. On both sides of the river is a cycle track, and then outside that are two car roads. This goes on for about three miles, and all the different types of transport work with each other, with bridges over the river and traffic lights for the boats and the cars. And it is habsolutely booful, with Dutch houses along the edges so that peoples can have a car parked in front of their house, and a boat parked on the river.

For a luffly boykin there was lots to watch and see. We had to use our motor for this part of our trip because of all the other bits of transport and the bridges, and as we chugged along, I saw ducks and children jumping into the river, and even a group of scarily hathletic senior ladies doing rowing, wot made Mum do a lot of finking about her wobbly bum, and wishing she was Dutch. Dad said if they retired to the Neverlands, he could have a boat in front of his house, and Mum could do rowing and not have a wobbly bum, and wouldn't that be marvellous ...

There was lots of stopping and waiting for the bridges to be opened along the river. At first, we all fort fings were a bit actual disorganised, and the men on the bikes who came along to open the bridges were a bit tardy and probababbly hanging around in the pub. Dad fort being a bridge man in the Neverlands could be an exerlent job for his retirement. After the third bridge, though, we realised there weren't 'bridge men,' but just one peddle-like-mad-to-get-to-the-next-bridge man. And no wonder we fort they all looked the same, just a bit redder and more puffed out the further we went along. Dad has changed his mind about retiring to the Neverlands.

August 8 (evening)
Tomorrow, the fuge ginger boyman and the Veggie-Tari-Man is taking the previously ginger one to Amsterdam. They say Mum is not to worry a-very-tall cos they has both binned there lots of times before, and will do exerlent lookering after the previously ginger one.

And also, there will be lots of rests and sitting down breaks cos Amsterdam has fundreds of coffee shops ...

August 9 (early)

I has learned a noo trick! I can do jumpering off the boat onto a pontoon. A pontoon is the wooden walkway across the water that peoples tie their boats to, and Dad says if Worzel can do it, can Mum stop being so Gerry-Hatrick about it.

I do fink Dad is being very actual hunfair to Mum. It's tooked me nearly a week to learn how to jump off the boat, and I honly worked it out cos Dad was having a hinteresting talk with some peoples on a boat a few berths down, and taking for-HEVER to come back.

In the Neverlands there is billions of boats, and in some places, instead of parking side on to the pontoons, boats get parked with either their nose or their bottom against the pontoon, and ... that isn't at all as easy as it sounds.

Our boat is 3.9 metres wide. That means, according to most Harbour Masters, we can fit into a box that is 4m wide. Now, Dad is a quite very actual brilliant sailor, but the rest of us range from totally incompertent to shambling hamateur. And that usually suits us all very fine cos Dad is so flipping fabumazing at sailing, none of the rest of us need to worry much at all. But gettering into one of them boxes needs skill and communercation and teamwork, and hideally, a bit more than 10cm to play with.

Yesterday, when we arrived at Goes-goose-house-however-you-say-it, and got hallocated our box, Dad decided that the best way – in fact, the honly way – to get into the space was pointing forwards. Wot meant cos he was at the back of the boat, he could do grabbing the bits of rope off the gnarly fuge boat-breaking bits of wood that make up the back corners of the space, wot you tie your boat to, to stop it floating away. And also cos we had a haudience of peoples waiting to watch the hidiotic Ingerlish peoples mess it all up, and Dad remembered that he was representing his country. Again.

So, we made it into the space and Dad gave himself a Gold Medal for not-making-a-plonker-of-himself. Because we is front-end-in, there is a big drop from the front of our boat to the pontoon. For a luffly boykin, this is not a fuge drop, hespecially as I saw Dad do it, and I has long legs and can clearly see wot I is jumping onto. So, I tried it. And I did it. And it was ... great!

Honly fings is not so great now as Mum reckons I will jump off the boat whenever I feel like it. She can't get off the boat cos her legs aren't long a-very-nuff, so could Dad stop yackering and grab Worzel before he bogs off and causes a Hinternational Hincident.

Dad reckons the honly person causing a Hinternational Hincident is the hysterical woman shrieking on the front of her boat, and if Worzel can get off, then so can she.

August 9 (later)

Dad has turned the boat round so that the very-much-lower back end of our boat is against the pontoon. It was either that or build some steps out of wood he doesn't have with him, on a pontoon he doesn't own, for a wife he Wants.

To. Strangle. None of which will get him any Gold Medals. From now on, if anyboddedy asks, our boat is 4.5 metres wide. Dad says.

August 10

We're going to an island today wot will be quite very peaceful and tranquil. Before that, though, we has got to go through a not actual peaceful-or-tranquil Biggest Lock in the Hole Wide World. Dad says that's hexaggerating: there's one in Panama that's a bit biggerer, but compared to the one we did the other day, this one is FUGE. And if the wind is blowing in the wrong direction it can be a Bit of A Challenge.

When the Veggie-Tari-Man and the fuge ginger boyman heard these words they did get quite actual halarmed. They crossed the North Sea in a storm wot Dad described as A Bit Bouncy, so they has worked out that Dad's Bits are everyboddedy else's STONKING.

August 10 (bedtime)

I dunno about your famberly, but my famberly makes the biggest, most redickerless noisy fusses about small fings. And they squabble about who should be making the tea, and sulk when someone pinches the last orange. But give them somefing difficult or frightening wot other peoples would have complete and hutter hystericals about ... and they does turn into Super Hooman Team Worzel's Famberly.

I are pleased to say that we did gettering through the fuge lock at Bruinisse magnifericently, but we was quite actual Ingerlish about it, so there was no high-fiving a-very-tall. At least not in public. There might have binned a teeny-tiny bit of that down in the cabin, and at least one celly-braterley cuppatea, but we did nuffink to let ourselves down. Heven I was perfick, even though there was four other actual dogs in the lock with us, and they all wanted to say hello luffly boykins.

I also learned another noo trick. I can now go up and down the stairs from the 'down below' bit to the 'up on deck' bit. I has gotted lots of confidence now, and my sea-legs are working so actual perfickly that I are moving about the boat without asking for permission or waiting for a lift. Mum says this is going to save a lot and a lot of heaving and lifting, but when we get to the island, we will be doing some re-arranging and barrier-finding.

August 11

There is horses on this island. I has no hidea how they got here, cos it is a tiny island, and there is no hobvious way on or off, unless you can fly or sail. As far as I can tell they is quite actual happy horses wot is wild, so they don't want to do saying hello to anyboddedy; not heven the previously ginger one. She was doing quite very well at getting close and being calm and quiet when the fuge ginger boyman appeared. And the fuge ginger boyman is fabumazing at lots and lots of fings, but quiet and hunobtrusive isn't any of them fings.

There is also lots of birds here, and I fink at one point there were a lot of crabs, but the birds must have dunned eating them all, cos all we saw were fundreds of crab shells, everywhere. It was a bit like a graveyard, to be quite actual honest, and I wasn't sure where I was very allowed to put my feets. So I tooked my feets into the water and did paddling in the shallows.

By the time I got back to the boat, the fuge ginger one had started to do a barbeque, but we had to very give up on that hole hidea of cookering outside quite actual quickerly because it bergan to rain, and then really, wheely rain and funder. When the lightning started, Dad had to give everyboddedy a great big long complercated hexplanation about why we weren't going to get strucked by lightning, and how we weren't going to die. I don't fink the Veggie-Tari-Man and Mum fort much of Dad's hexplanation, because they decided the honly sure way to havoid being struck by lightning and very-not-getting-Cabin Fever was to drink some wine. Lots of wine.

August 12

According to the previously ginger one, all the windows in her bit of the boat leak. Dad says this is not true and the wet sogginess there is cos we had all the windows shut last night because of the rain, and five peoples and a dog breathing all night has made somefing called condy's-nation, wot is when all the water falls out of people's mouths and onto anyfing they don't want to get wet.

You will not believe how much condy's-nation we all made last night. It was a-very-nuff to completely soak the previously ginger one's duvet, hand half her clothes. But the windows are NOT leaking, Dad says.

August 13

It rained again last night. The previously ginger one's bed is now wet through to the mattress, and Dad says the windows are *still* not leaking. We have decided to give up on the hole island-living-back-to-basics hidea cos back-to-basics needs sunshine and a light breeze, and not rain. And also the solar panel wot makes the fridge stay cold to work.

We is going to head to Ouddorp wot is Dad's favourite actual place in the hole of the Neverlands, so we can stop making so much condy's-nation by doing all our breathing in the Yacht Club bar. And hopefully, we will be able to dry out some of the previously ginger one's bunk, cos she's planning on nicking Dad's bunk tonight, and then he'll sleep on the bench seat and snore all blinking night. And keep everyboddedy awake. But at least if he does that fing, we will all be awake to watch the windows not leaking.

August 14

Ouddorp is my favouritist place we has visited so far, mainly cos the peoples that run the Yacht Club has got a Labrador called Pepper. At first, Pepper wasn't too actual sure if luffly Ingerlish boykins was allowed in her Yacht Club, but after some Hinternational Diplomacy, and some bits of chicken and biscuits, I are pleased to say that me and Pepper are hofficially friends.

Being hofficially friends with Pepper has meant I has had some off-lead zoomie times, and I has binned edercating the hole of the Neverlands about being a Lurcher. It seems that Lurchers are quite not actual common here, and it has binned my himportant work to show everyboddedy wot one is. And convince them that I aren't hunderfed. Mum says if One. More. Person asks why I are so skinny, she's going to go bonkers.

I ARE A LURCHER

I are a Lurcher,
I are s'posed to be lean
I aren't skinny, or half-starved
My Mum isn't mean

I are a Sighthound
My ribs they should show
I get plenty to eat
I fort you should know

I are a long dog
I run fast and hard
Wot I couldn't do
Covered in lard

I are a Lurcher
I don't need more food
Please stop saying that fing
It's quite actual rude!

Honly to the Dutch peoples, it isn't rude. Dutch peoples be very quite actual straight-talking. And if they does want to know somefing, they get straight to the point. They don't bother with wiggling round a subject before they get to their questions, and that can be quite actual surprising for Ingerlish peoples. The good fing about all this is that you don't have to wonder about whether they is keen to be your friend, or just being plite. If they like you, they is very fusey-tastic about that fing, and they are kind and generous, and generally speak perfick Ingerlish.

August 15
Our neighbours in Ouddorp are a tiny, hexcitababble, cheerful-terrier-muddle, and a quite very cross Bulldog. Every time I go up on deck, he does shoutering at me. It doesn't last long, though, cos the poor grumpy dog can't do breathing very well. It is quite actual sad, really: he gets himself all worked up about this-is-my-harbour and this-is-my-boat, and is just about to get to the hinteresting fings he wants to tell you ... but then he has to have a bit of a cough, and do some breathing out of his mouth. Then the cheerful-terrier-muddle skitters around with his cheerful yapping wot must really, wheely annoy the Bulldog, wot never gets to finish telling everyboddedy his forts cos he can't breathe.
Maybe that's why he's so grumpy.

August 16
I has dunned bashing my shins, and now I feel like a proper sailor boykin. I has got two bruises, and one even has a bit of scrape on it, but it is very nuffink to worry about, and I has declined Mum's kind offer of some Sudocrem and clucking. Noboddedy else has got clucked at or had to have Sudocrem on their bashed shins, and I can tell you that, by now, all of us is wearing lots of leg bruises. It's wot happens when you do sailing.
I did not get my shin bruises doing sailing. I got them leapering off the boat to go and see Pepper. It's my own actual fault, Dad says, for not doing

'wait,' as in wait-until-I've-pulled-the-boat-a-bit-closer. And trying to leap eight feet off a skiddy deck surface.

We're leaving Ouddorp tomorrow, and mostly we will be quite actual sad about this. It has binned fabumazing here: very quite relaxing, and so booful. I shall be sad to say goodbye to Pepper, but Mum says we will always visit Ouddorp when we come to the Neverlands, and I will definitely see Pepper again sometime.

Dad says he is quite actual glad we is leaving tomorrow, because someboddedy on our boat has dunned a bad and hembarrassing fing. And there is hevidence as well. Someboddedy has dunned a POO in our LOO wot is very very Not Allowed, hespecially when we is in harbour. Cos the poo has comed out of our boat and now it is Floating in the Water, bobbing Right. Next. To. The. Boat. And anyboddedy who comes to visit us will see it. Dad's tried giving it a bit of a poke with a stick to get it to sink, but it won't go away.

Dad would like to do some hinvestigating and discussing who did this terribibble fing, but everyboddedy in my famberly is in the Nile about it, and I don't fink that will change, however much Dad grumbles. All of our Dutch neighbours do speak perfick Ingerlish, and if he starts to hinterrogate my famberly, all he is going to get is a lot of 'it-wasn't-meeee!' talk, that will draw even more very hattention to the poo wot is still bobbing Right. Next. To. The. Boat.

I can tell you for nuffink that 'it-wasn't-meeee!' but I does know hexactly who it was. Mum says it was a haccident, and somefing to do with too much coffee, and I are to remember that I has binned Swored to Secrecy and very actual bribed with bitsa cheese and chicken wings ...

August 17

Today, we will be doing our biggest journey since I has been on board, sailing from Ooudorp, all the way down the Grevelingenmeer, and back to Zeirekzee. Mum wants to check that the car is still where she lefted it, and also, I has discovered, we has Friends in the Neverlands who we will be visiting tomorrow.

Late last night, Mum did confessing about the POO in our LOO, and did try to hexplain it wasn't somefing she had planned to happen. And although Dad has dropped all the POO in our LOO talk, he says today he is going to put up the Spinny-cure. And then he looked at Mum, and dared her to do complaining.

The Spinny-cure, I has discovered, is the biggest sail on our boat. And, hunlike our other sails, it is all different colours, and isn't triangle-shaped. It's like a fuge shield, and when it fills with wind it makes the boat move really fast.

It is attached to a fuge pole that wobbles and bangs and crashes about making horrid noises wot Mum doesn't like. Mum says we is both going to have to be brave about the horrid noises, cos otherwise Dad will tie the Spinny-cure pole to the boat and that makes the boat heven more wobbly.

I do not know why my Mum has such hystericals about the Spinny-cure, to be quite actual honest. It is a bit flappy and clangy, but that is all very out-waved by other fings, like zipping along and beating all the other boats in a race they don't know they is in. The fuge ginger boyman and the Veggie-Tari-

Man have spended all day spotting a boat to chase, and then been working really, wheely hard to catch it up and overtake it. The previously ginger one has not binned on Mum's side, neither. She loves the Spinny-cure sailing cos the wind doesn't blow about in the cockpit, and everyfing feels warm.

Dad has spended most of the day tutting at people who are either in motorboats or who haven't got their sails up correctly. Mum reckons that's cos those men have got their wives onboard, and they still like their wives and want to do keeping the peace. Then she stomped off down into the cabin to have a snooze, and pretend the hole Spinny-cure fing wasn't happening.

August 17 (later)

Spinnakers can honly be used when the wind is behind us. It is complercated, but does mean that Spinny-cure flying doesn't last that long. When it was safely putted away in its bag (for-Hever, Mum hopes), the Veggie-Tari-Man decided that he was going to pretend to be the pretty lady in Titanic, and balance right at the front of the boat Not. Holding. On. Mum decided that he was a Big Boy and could do killering himself without her hinterfering.

Not to be outdunned, the fuge ginger boyman decided he would do standing on the edge of the boat dangling from a bit of wire, tied to a bit of rope wrapped through the belt holders on his jeans. Then he did lean out like he was going to walk backwards down the side of a mountain. Mum worked quite very hard at trying to remember that he also is a Big Boy, but do-you-have a-very-nuff-sun-cream-on? and are-you-sure-that-rope-can-take-your weight? Dad worked quite actual hard at Not Dunking the fuge ginger boyman in the water. He was quite actual tempted but he said it would slow down the boat ...

August 18 (early)

We're off to a party today, and I are very quite hexcited. Mum says there will a garden and, most himportantly, there will be grass wot I can lie on, and wriggle about on, and feel some luffly, cool, earthy, land type of stuff hunder my feets for a hole afternoon.

It seems that Mum should have dunned less remembering that the Veggie-Tari-Man is a Big Boy when he pretended to be the pretty lady from the Titanic yesterday. This morning he has woked up with bright red feet, and he can't put on his shoes. The fuge ginger boyman can put on his shoes, just very fine, fanking oo kindly, but there is other bits of his boddedy that seem to have got very actual squashed when he was dangling on the wire, so he is doing a very good himpression of walking-like-a-duck.

Me, Dad, and Mum have lefted the previously ginger one in charge of the walking wounded and stoopid. She is to make sure that the Veggie-Tari-Man keeps his feet out of the sun, and is to phone Dad himmediately if the fuge ginger boyman even finks about moving the boat.

August 18 (later)

The party was fabumazing, and there were trees to sniff and cool grass to lie on, and lots of friendly peoples who all wanted to say hello to me in a gentle and fortful way. It was Chantal's birfday and although we do know Chantal from Ingerland, she is actual Dutch, and because her birfday was one of them with a zero at the end, she had decided to go home and have her birfday with her famberly.

There was coffee and cake to start with. Apparently, that is quite actual traditional in the Neverlands, and after that there was all sorts of party food and drinks and nibbles. And Spike.

Spike was not for dinner. Spike was a little Chihuahua and Maltese-cross laptop dog. And I have got to say I has never met such a hinsistent, fusey-tastic dog. I does also fink he had some seriously fuge problems with his memories.

When we first did meeting, we did all the hello-luffly-boykin sniffing and smelling stuff wot is quite very normal and happropriate. But it seems that Spike kept forgetting that we had already dunned getting-to-know-you sniffs cos every five minutes he wanted to do it all again. And again. And again. In the end, I did lying down, and tucked all my pleased-to-meet-you smells hunderneath me.

Spike is honly a teeny tiny dog, but every time I did a pee on somefing, he did a pee on that same fing. I fort he would run out of pee after a very while, cos it should not be actual possibibble for one tiny doggy to have that much pee inside him. But it is. Apparently. I runned out long before he very actual did, wot I was a very bit disappointed about, so I just did pretendering-I-was-peeing, wot seemed to trick him quite very nicely.

I would not like anybboddy to fink that I did not like Spike. There was lots of luffly fings about him, and he was not unkind to me, and we did some exerlent playing as-very-well. But if dog peeing and sniffing is ever hintroduced in the Holympics, Ingerland is going to have to find a betterer competitor than me. Otherwise the Neverlands will definitely get the Gold Medal.

August 19

We're going to Willemstad today, and it is the last place we will visit in the Neverlands. Mum says we have to leave visiting Willemstad until very last because, if we went there first, we wouldn't go anywhere else.

Willemstad is an old town with a moat, wot is in the shape of a star. It has an outer harbour in the moat, and an inner harbour. It is my famberly's plan to get our boat into the inner harbour, wot is quite actual near all the shops and nightlife, but it is halmost quite very himpossibibble to achieve.

It's like trying to park in Southwold in August, Dad says, and there might be some quite very long waiting about, and wriggling and moving the boat around, but we will get into the inner harbour if it's the very quite actual last fing we do.

I fink Dad's forgotted he's got Mum onboard. Either that or he doesn't know about her special and weird and quite very fabumazing Parking Karma. Mum reckons I might ruin it all now by telling everybboddy about it, but I fink you should know that Mum has Magic Powers. She's quite actual rubbish at remembering wot bin colour should be put out, and is very clumsy and worries about actual everyfing. But not about getting a parking space. And if there isn't a space, somehow, just as Mum arrives, someone will come back to their car and discover they has got a fuge, ignormous hurgent need to move it so that Mum can park.

Mum's not sure if it works with boats: it worked in Zeirekzee, but she's not sure whether it was her Parking Karma or the Harbour Master realising Dad knew how to tie ropes, and wanting someboddedy sensibibble by the wall.

So she says we will Wait-and-See. And not say anyfing in case it doesn't work and everyboddedy laughs at her.

August 19 (later)

We are in the Inner Harbour, and Mum is hinsistering it is all because of her Parking Karma. I aren't sure it works for convincing peoples-you-haven't-binned-drinking-too-very-much-wine, though. Or for convincing Dad you isn't Away with the Fairies. Dad finks we has a place in the Inner Harbour cos of Luck, but Mum is sticking to her Parking Karma feery. And the previously ginger one has been very quite loyal (here-have-some-moneys-for-an-I-Scream), and has told Dad that Mum *does* have Parking Karma, cos she's been in a car plenty of times with Mum and seen it happen. The fuge ginger boyman doesn't believe in Luck or Karma or anyfing that can't be proved by Science, so he has decided to join Dad's side, and now everyfing is quite very actual noisy.

Very actual sensibibbly, the Veggie-Tari-Man has decided that the safest way to stop the hinter-generational-arguing-for-the-sake-of-it is to suggest we Murder Someboddedy. Wot my hentire famberly finks is a fabumazing hidea and wants to know how to do it.

Dad says that all murders will have to wait until this evening, though, cos we have got to get the boat looking very quite booful and ship-shape: his little brother is coming to visit tonight, and he wants our boat looking perfick for when he does. And he finks he'll want to do murdering as well.

Dad doesn't have a brother, so now I are completely confuddled. Maybe he's the one who has binned drinking too much wine with the fairies.

August 19 (evening)

Dad's 'brother' is called Steven, and he has a wife called Nastasha. He isn't really his brother: Dad dopped him very hunofficially about ten years ago as his Little Dutch Brother. I are almost sure that Steven finks this is okay, heven though, like every other man in the Neverlands, he is 40cm taller than Dad, and has an himportant job as a Hengineer. Nastasha, Steven's wife, finks it is quite very hysterical, and like Mum, is just very actual glad she can have a few hours off from listening to her husband yacker on and on about the hendless, hunimportant will-they-ever-shut-up details of sailing.

Heventually, Dad remembered that we was all supposed to be murdering someboddedy, but it turned out that cos this was the Veggie-Tari-Man's game, it was all Veggie-Tari-Man, and there would be no blood. I don't fink anyboddedy was that disappointed about the No Blood fing cos there was a lot of lying and cheating and trying to prove people's wrong. I would like to do hexplaining the rules but I can't; they is just too actual complercated, and I are quite very certain they changed halfway through cos Dad is far, far too blinking good at lying and cheating. Or else everyboddedy else dranked far too much wine.

August 20 (first fing)

Today it is our mission to find a vet. There is very nuffink wrong with me, fanking oo kindly, but it is all to do with the poxy Guv'Ment, and me not going to prison.

We does know there is a vet in Willemstad, because we looked on the interweb, but Mum is still in a right actual flap about it all because we still have to find it in the town, and I have to eat a tapeworm tablet before I can go home. And wot if Worzel won't eat the tablet or his microchip doesn't work or he refuses to be fiddled with, or the vet doesn't know what a Lurcher is, and finks Mum is mean for not feeding me a-very-nuff.

Dad is coming with us to the vet.

August 21

The vet was a luffly, crinkly, wrinkly-faced man who had dunned lots of dealing with hysterical Ingerlish mums before. And I are pleased to say he did know hexactly wot sort of doggy I are, and didn't say nuffink about me being skinny or hunder-fed. I was an exerlent boykin about everyfing at the vet, and Mum says she can half do relaxing now. She was really, wheely worried about the hole fing, and panicking about the paperwork being correct.

Dad tried to ask her if she felt better now, but she tolded him that she will honly really stop worrying once she is back in Ingerland with Worzel, and she knows I won't want to be fiddled with by a strange man in uniform who needs to scan me, and then I get put in prison, and Dad won't be with her. And she'll be so nervous people will fink she has got somefing to hide, and what if they search the car for ill-eagle himmigrants and Worzel gets frightened ...?

The peoples on the boat next door do come from North Yorkshire, wot means they support Leeds United Football Club. Dad has had a Very Pleasant Hour talking to them about their life in Leeds, and mining and teaching and making wires, and allsorts. To be actual very honest, they could have comed from Mars and have a passion for inter-galactic tiddlywinks and he would have henjoyed talking to them. Anyfing, he says, is better than listening to Mum wind herself up into a gibbering wreck panicking about blinking Worzel-What-Ifs.

August 22

I are safely back in Ingerland, and all Mum's worries and wobbles about me finking I was being attacked with a scanner by a strange man in a uniform camed to actual nuffink.

When we got to the barrier in our car, the lady in the ticket booth handed the scanner to Mum so that she could press the button whilst I was still strapped into my car seat belt. The number came up perfickly, and Mum handed back the scanner, droved round the corner, and letted all the air out of her mouth wot she had helded onto since last night, all in one fuge gulpy rush.

After that, everyfing was quite very simple, and the same as it was on the way to the Neverlands. I slept in the Fer-eee Prison but for the last hour, and

Mum tooked me to the Dog Deck to watch us arriving back in Ingerland. I did decline her kind offer of a wee on the deck. You would fink she would have got the hidea that I aren't weeing on any deck, no matter how big it is. But I did henjoy watching the water tumble about when we arrived in Harwich, and I was very quite pleased to say hello to the other luffly dogs waiting up there.

Tonight, I has binned to have some proper zoomie mad times with Kite, who I missed a lot and a lot. I know I did miss her grass and being able to run about in a place I feel safe and relaxed in, and that is halmost the same fing I do fink.

At home, all the cats have dunned coming back to say hello, although Gipsy has binned A Cow, according to Kite's Mum, by not coming in with the others at dinnertime. She says she had to trick her by coming round at all different times of the day, wot Gipsy did not happreciate a-very-tall. She likes to be helusive and not helpful, and definitely Not Tricked.

Our garden has gone bonkers whilst we was away, and although our neighbours did lots of picking the tomatoes and runner beans, we don't fink any of them like courgettes. Mum says when she has unpacked the car and started the washing machine going, her next most hurgent job is going to be finding Hinteresting Fings to do with Marrows. About 17 of them.

August 24
Dear Veggie-Tari-Man

I are very actual sorry I did not get to say goodbye to you before you did returning to the North. I hope you did henjoy your holibobs with us, and Mum says you is welcome on our next trip away if you can face the fort of being stucked with Us Lot again.

Since you went home there has binned a lot and a lot of gee-tar playing here. Dad and the fuge ginger boyman fink they need to do some more practicing if they is hever going to be as good as you. Fank oo for all your singing and playing on our boat. You did make our holibobs very quite special, and we didn't have to do any disagreeing about wot to put on the radio. We just putted you on instead.

Dad says you had a better trip back to Ingerland than you did going to the Neverlands. Mum says you can come in the car on the ferry next time if you do prefer, but maybe a big bouncy gale is preferababble to Mum and the previously ginger one shouting at Mr Sat Nav?

You will be pleased to actual know that since we did getting back from the Neverlands, Dad is still hinsistering that the windows don't leak, although the hole in the deck right above the previously ginger one's bed probababbly, possibibbly does.

From your luffly boykin

Worzel Wooface

Pee-Ess: Sorry about the hole fridge full of raw meat that you had to grope around every time you wanted to find the butter or some milk. I has heard that since you camed back from our trip, you has becomed a vegan, and I are worried this is All My Fault. Chicken wings is very actual crunchy, and there is not a fuge amount I can do about that, but I does happreciate you might not want to hear that first fing in the morning. Or step on one I has forgotted about. With Bare Feet

August 26
Now that I has been home for a few days, I has had time to do finking about My Sailing Holibob in the Neverlands.

- Peoples in the Neverlands are not called Neverlandsish. They is called Dutch, wot makes no sense a-very-tall
- 'Do you speak Ingerlish' is a very quite plite fing to say, but also very actual pointless. Everyboddedy there over the age of 12 can speak better Ingerlish than most Ingerlish people
- Dogs is quite very pop-oo-la in the Neverlands. Picking up dog poo seems to be heven more unpop-oo-la than it is in Ingerland, though, wot we did find very actual surprising
- In the Neverlands, all cafes and restaurants are very welcoming to dogs. We did not find one single place that was not pleased to see Worzel Wooface, and I was offered plenty of space for my bed, hand a drink of water
- Everyfing in the Neverlands is actual quite flat. This means that peoples who hasn't dunned riding a bicycle for twenty-five years suddenly fink they might like to have a go. With Worzel in tow. Heventually, they do realise that they should do handing the dog over to someboddedy who has some more recent hexperience of bike-riding. And who is also young a-very-nuff to promise not to fall off, or in any way hurt, frighten or, most himportantly, let go of Worzel
- Bikes in the Neverlands do not have brakes. To stop you has to pedal backwards. No Worzels was hurt in the process of discovering this fing. One Mum was, though
- There is special tools that can help with the hole rope-hooking fing in locks. Dad says these are for Bad Sailors. Mum finks she qually-fries for one
- Dad is a really, wheely fabumazing sailor. We fort we knewed this before, but when a lock man tells fifty other boats to 'do-it-like-he-did-it' in a crosswind with the lock water already starting to churn, you does realise just how good he is
- Previously ginger ones who want to do jumping off boats to go for a swim should probababbly work out how they will get back on again. Before they jump off
- Noboddedy in the Neverlands has ever heard of cider. No matter wot accent you say it in, or how many times you try to explain wot it is

September

September 1
****HIMPORTANT DISCOVERY****

I are betterer at football than Kite! Finally, there is somefing wot I can do betterer than she can. That's not saying much, to be quite very truthful, cos Kite is actual rubbish at football. All she wants to do with balls is pick them up. Even when they is the size of her head, and they won't fit in her mouth. How-very-ever, I does not have this actual hobsession, and I will hoccasionally do giving a football a biff with my paw. And then Mum squeals Good BOY, Whizzy Woo! I can see we is going to be doing more of this football malarky over the next few days. Well, until Kite finally manages to sink her teef into the ball ...

September 2

Wot is it with hoomans about peeing outside? For some reason when a hooman gets to four years old, it does become hunacceptababble to do weeing anywhere but in the loo, and I does find this actual quite hard to hunderstand.

Happarently, when you is a 46-year-old-Mum in a reasonababbly smart outfit , driving to a party in the middle of no-actual-where, and Mr Sat Nav is sending you round and round in circles, through posh villages, or lanes where there is no hedges or quiet country lay-bys with convenient places where you can hide up, it is not hacceptababble to stop the car, do a wee, and get back in the car. And stop jiggling and making sobbing noises and finking that soon you isn't going to be able to go to the party cos your trousers will be all soggy. And then realising that you Don't. Just. Want. A. Wee. And finking that even if you find somewhere, you probababbly won't be able to do just a wee, and you cleaned out the car yesterday, so the honly fing in it now is a bottle of diet cola wot the previously ginger one had left behind, and a Worzel.

There does come a time in every dog's life when a hooman will do somefing that has to be staying between Them And Their Dog. Happarently, hoomans have a saying that is about keeping somefing between Them and Their God. Hobviously, someboddedy has spelled that all wrong cos I are convinced that if this God person had dunned seeing wot Mum did, they would not have binned able to actual resist tellering everyboddedy, but because I are A Dog, I shall not be tellering anyboddedy. Apart from saying that diet cola isn't at all very sticky and it is flipping, blinking hunbelievababble wot a Mum can do when she truly, deeply, habsolutely has no choice, and she is nearly almost, completely, probababbly I-don't-care-anymore certain that noboddedy is watching.

And we will Not. Be. Speaking. Of. This. Again.

September 3

After all that quite actual fuss getting to the party, and the hole not-speaking-of-this-again hincident, I did fink that I was dunned with hexcitement for one day. But there was a noo hexperience at the party wot I are very quite keen to do remembering about.

Nog Roast. I would like to actual know why I has got to very nearly actual Three. Years. Old, and noboddedy has fort it himportant to hintroduce me to a Nog Roast before. I does feel, now I has hexperienced Nog Roast, that I has binned Neglected, and it is very quite Hunfair that I has missed out on three hole blinking years of Nog Roast, simply cos I did not know they hexisted.

The bestest fing about Nog Roast is that there is always far too actual much for all the peoples to eat. And cos it smells fabumazing, hoomans get very quite fusey-tastic about how much they fink they can fit insides themselves. Then the hoomans do all feel a bit actual hembarrassed about how much they has got lefted over, and worried that they will get tolded they has hiballs-bigger-than-their-bellies.

I are halmost actual certain that this is why I was hinvited to the party, and why everyboddedy was so actual pleased to see me. It was my himportant work to do helping peoples get rid of the hiballs-bigger-than-their-bellies hevidence. You will be quite very pleased to know I did do my bestest, don't-tell-anyboddedy-Worzel gentlest and quietest taking of all the bits I was offered.

How-very-ever, I would like it to be actual knowed that I are not a dustbin, nor is I a hog-pig Labrador, and I did plitely decline all the bits of yucky white bread that some of the smaller hoomans fort I might like to do disappearing for very them. And also their tommy-tar-toes. I fink they was a bit disappointed that I would not do eating this stuff wot their mum's had putted on their plates, and there were a lot of but-I-want-trifle, and not-until-you've-finished-your-dinner. And when someboddedy did wonder if I would like a grape, Mum was very quite quick to say no-fank-oo-grapes-is-poisonous-to-dogs on my actual behalf.

Wot is a good job, I do fink. After all the dramaticals and hystericals and Fings. We. Will. Not. Speak. Of. Again on the way to the party, I does not fink Mum could have dunned coping with the same fing on the way back home. But it might have gived Mum somefing betterer to say than the hawkward fib she camed up with when the previously ginger one asked wot had happened to her bottle of diet cola ...

September 4

The man who will be driving the previously ginger one to college each day this year has just comed round to visit, and hintroduce himself to the previously ginger one, so that she is not worrying about who will be taking her. They had a good very actual chat, and the previously ginger one did say she likes to listen to music through her headphones, and sit in the back and Not Do Talking. And sorry if he finks she is being rude, but it helps her if she can listen to her music, and not panic about wot it will be like when she gets to college.

Mum's ever so very himpressed. If honly everyboddedy was so considerate to people who have a Ment-all-elf, life would be a lot and a lot easier. Now the previously ginger one won't be worrying before she goes to college, and she won't arrive at college having been hexpected to do all sorts of hunnecessary chattering and being plite. She'll be a bit more relaxed, and then all the peoples at college will have a better day, too.

And the taxi man might heven get a smile from the previously ginger one first fing in the morning, which is more than Dad gets. But that's probababbly got nuffink to do with the Ment-all-elf, and more to do with the fact that it's seven o'clock in the morning, wot neither of them is any actual good at.

September 5

Today has binned fabumazing. Me and Kite found somefing small and smelly to roll in. Neither of us was quite very sure wot it was, but we tooked exerlent turns at rolling in it, and watching each other rolling in it, and also trying to stop Mum and Kite's Mum hinvestigating wot it was that we was rolling in.

Mum says that sometimes she does wish her life was as simple as mine ...

September 6

For a quite actual small moment today, I was the most very cleverest dog in the hole wide world. I was! I really, wheely was.

I stooded in poo. That bit wasn't that clever, to be quite actual honest, and noboddedy is quite very sure whose poo it was, but it got stucked to my paw. Then I did jump in my paddling pool. Mum fort I had jumped in there to do washing it off, and was super-himpressed. She started saying fings about me being hintelligent, and not-as-daft-as-I-look, but I ruined it all by lying down in the paddling pool. Now, according to Mum, I are the most stoopid dog in the world. And also Dad's dog.

No, she doesn't want to be leaned on.

September 7

HAPPY BIRFDAY SONG

Happy Birfday to my friend, Kite
She's a Lab, but she's alright
We run and chase, and do play-fight
And tug on toys with all our might

Happy Birfday to the Girl Next Door
Who always seems to lie on the floor
And waggle her legs, to make quite sure
I'm coming back to play some more

Happy Birfday to my Partner-in-Crime
You're four now, so you're in your prime
I luffs you mostly all the time
Hespecially when you're covered in grime

Happy Birfday to my pal, Kite
Who shares her garden, day and night
And her balls and toys, that's right
We're bestest pals; it was luff at first sight

September 8

I has got a Complaint to the Management, and it is all about my Not-Grass-Really-Carpet. I don't fink Mum and Dad did do any of the hole finking-it-through-proper-like, as it's under the napple tree cos nuffink will grow hunder there. Plenty of fings have grown *above* it, though: bllions of the blinking fings, and if one more napple decides it is going to hunexpectedly fall off the tree and land near my head, I are going to have a Serious Sulk. It's like being in a war zone.

Just when I are breathing a sigh of relief that the napple bomb didn't hit me, the second wave starts. Cos these isn't normal bombs, they is bi-yo!-logically weapons, full of angry, I've-just-falled-8-feet-and-landed-with-a-thud hinebriated wasps, wot come steaming out of the napple like a bunch of drunks on a Saturday night, spoiling for a fight with whoever just hoffended them.

And the first fing they see is me, trying to henjoy a bit of late summer sun.

I would like the wasps to know that I are a hinnocent by-snoozer in all of this. Mum planted the tree, the fuge ginger boyman didn't let Mum prune it, and Dad is too busy with his boat (and too hinterested in a quite very easy life) to chop it down. So I are going to stay hindoors and away from my Not-Grass-Really-Carpet until the danger has passed. Someboddedy else can get bombed. And stung.

September 9
****FUGE NOOS****

All of a quite very actual sudden, I has got a new foster brother. His name is Harry, he is 13 months old, and he is a Lurcher, and although he has black fur, everyboddedy finks he is Just Like Me, cos he's skinny and leggy and pointy, with silky fur on his ears.

Harry isn't like my usual foster brothers or sisters because he does have a Dad Wot Luffs Him, but because Harry's Dad losted his job, and then his house, and was sleeping in his car, he was starting to fink he was going to lose his dog, too. And then the very will to live. He could actual cope about the house and the job, but the fort of losing Harry was too actual much to bear. So Hounds First stepped in, and we are his noo foster carers. Harry, that is, not the Dad Wot Luffs Him.

When Harry and his Dad camed to visit us today, there was lots of talk about Harry being safe, and not to worry. Harry's Dad Wot Luffs Him was worried, though, you could tell. He's never been without Harry, and he didn't want to leave him.

So Mum did the honly fing she could fink of, wot she always does when peoples are sad and wobbly: she got really, wheely bossy. And told Harry's Dad to remember that Harry is His Dog and Not Our Dog, and to go and sort out everyfing else in his life. .. and then Come Back. But not to come back until he was ready to take Harry home cos that would hupset both of them.

Dad stayed in the Office when all this talking was going on. I aren't sure if he was hiding cos Mum was being bossy and hembarrassing, or whether it was cos he fort Harry's Dad Wot Luffs him would start to do crying, and Dad would have to do joining in.

September 10

Harry is Not. Like. Me.

Harry is noisy and hexcitababble, and very fusey-tastic about actual everyfing. Mum says he is honly a baby, heven though he is nearly the same height as me. And we are to be tolly-rant ,and remember he has spended the last eight weeks living in a car, and is probababbly, definitely missing his Dad.

He might be missing his Dad but I are missing my peace and quiet, and if he doesn't stop barking at me All. The. Time, I shall soon go bonkers-crazy.

September 12
I tolded Harry off today. Not Big. Time. Badly. Just a bit of a will-you-behave-and-stop-poking-me. And do you know actual wot? It worked. Harry did stop barking at me, and then he did exerlent sorry-Mr-Wooface wiggles, and now everyfing is quite very actual peaceful.

Mum is very himpressed with me. And so is Dad, now, after Mum hexplained that my telling off was somefing called pro-porch-in-all, and it didn't mean Worzel was being nasty or no longer a gentle boykin. Good doggy manners isn't just about being luffly to every dog you does meet; sometimes, a more senior doggy has to ask babies to behave themselves if they don't notice that the senior doggy has had a-very-nuff. And so long as it is pro-porch-in-all, it is a good fing. And quieter. Much, much quieter.

September 14
Now that me and Harry has reached a hunderstanding about Do Not Bark in Worzel's face, and Do Not Poke Worzel when he is asleep, we has started to become friends.

Harry knows all the rules of bitey-facey, and does play nicely and not too actual rough. He knows all the right noises, HAND when to make them. And when to stop and start again. We is like a perfick dancing partnership, and I does fink it is fabumazing.

Dad doesn't. Well, he does, but he'd rather we didn't do bitey-facey singing and dancing by his hearholes. On the bed. On a Sunday morning. But I fink once Harry removed his foot from Dad's mouth, Dad did quite henjoy watching us whilst he drank his cuppatea.

So very actual much, actually, that, at one point, he did start to join in with making bitey-facey singing noises. Then Harry and me did give Dad the hexact same wot-are-you-doing-you-stoopid-man look that Mum gives him when he's had too very much cider, and he's trying to make himself some dinner. Wivvering, Mum says it's called. Dad got up in a sulk, cos he felt like he'd been hoomiliated by a pair of Lurchers, and bogged off down the boat.

Mum's not so sure how she feels about this; she fort only she had special make-Dad-feel-like-a-six-year-old powers, but if a pair of Lurchers can do it, then maybe they're not so special, after all.

And maybe she should be gentler and kinder to Dad.

September 15
There was a secret meeting tonight, and Mum has been quite actual silly, going on about cloaks and daggers. I do fink she should stop all this dressing up talk ,and take this all Far. More. Seriously.

The meeting was all about wot to do about the Cow Sell and its stoopid beach ban. A cunning plan is being made, happarenty, wot Mum can't say nuffink about cos if anyboddedy asks, she wasn't there.

She was, though! I does know she was cos she came back smelling of Auntie Charlotte and Ninja-the-Terry-fried-Terrorist. So she can pretend to the hoomans as much as she likes but she can't fool Worzel Wooface.

Mum's been giving Aunty Charlotte and other peoples wot like dogs and the beach some polly-trickle hadvices, but if the Cow Sell finds out Mum's been helping Auntie Charlotte, it might make more troubles than it's worth.

Anyway, Auntie Charlotte and her friends has decided that they is going to make a proper, hofficial group to protect the rights of dogs, and then the Cow Sellers are going to have to listen to them. And if they doesn't, one of the peoples from the group is going to haccidentally drop bits of paper in front of the poxy Cow Seller for Southwold, pretendering that they will be standing in an helection against him. And if that doesn't work, Auntie Charlotte is going to actual HAVE to stand in an helection against him ... Now Auntie Charlotte is having to drink wine and have a lie down cos she really, wheely doesn't want to stand in an helection and become a poxy Cow Seller. But she will if she has to. And if there is plenty more wine ...

September 17

Mum says that seeing as Harry will be here as long as his Dad-Wot-Luffs-him does need us to look after him, Harry is going to have to learn the House Rules. To be quite actual honest, they isn't the House Rules a-very-tall, they is Mum's Rules.

Dad has a few rules, but they is mainly about light switches and cupboard doors and ketchup, so they can mainly be completely hignored by luffly boykins like they is by everyboddedy else.

****THE QUITE VERY ACTUAL MOST HIMPORTANT HOUSE RULE****

- 🐾 Do not bark at the cats in the garden. Fings will get loud, and then they will get painful. For you. The cats can go hunder the fence. You is not a Wizard, the fence is not on platform nine-and-three-quarters, and you will biff into it, wobble the napple tree, and make all the napples fall off the tree on top of your head
- 🐾 Do not chase the cats in the hall. The hall floor is skiddy, and you has rubbish brakes. The cats will go through the cat flap, then turn around to wait until you is outta control, and biff wotever bit of you ends up jammed in the cat flap
- 🐾 Do not chase the cats in the kitchen. You will fink you is chasing one cat, and will not notice that the other four are also in there. Five against actual one isn't going to end well. And heven though most of them don't speak to each other, Mum says it's somefing to do with Blood is Ficker than Water. You is Fick if you do try to take them on, and they will draw Blood. Yours
- 🐾 Do not chase the cats anywhere else, neither. Somefing similar to all of the above will happen if you do do this fing

I fink there is probababbly some fings missing from this list. But, currently, Mum says, these are the honly rules wot matter, and until Harry gets these rules, then Nothing Else Matters.

THREE QUITE ✓ *very* actual CHEERS FOR **Worzel Wooface**

September 18

Our cats have dunned lots of fostering of dogs before, and they is usually quite actual good Judges of Character and communercating their forts to Mum. If there is even the slightest whiff that a foster dog might fink cats are Dinner, they go and hide in the shed and stay there. And there is nuffink that will change their minds; not heven Mum with the bestest treats in the world.

Harry doesn't want to eat the cats or hurt them, but he is very actual over-hexcited and redickerless around them. His cunning plan to get the cats to hunderstand him means he keeps barking louderer and louderer at them, just like Mum when she tried to speak Dutch. They didn't hunderstand her the first time; they didn't hunderstand her any better when she said it louder, and please could she go away now, cos she's frightening all the other customers?

Everyfing between Harry and the cats is getting lost in translation. The hisses and the clattering of the cat flap, and Harry's hendless barking and chasing the cats is not helping anyboddedy – most of all me. I has found myself wanting to join in with all the hysterical woofing and chasing, and I has binned on the wrong end of a couple of sharp 'NO!' words from Mum. Even Gandhi, who is supposed to be an Onorary Lurcher, and my bestest pal, tried to take a swipe at me yesterday ...

Dad says this is how wars start.

September 19

I fink I has worked out why the Queen has all those posh dinners at Duck-and-Ham Palace. It's a quite very actual fact that it is himpossibibble to shout and eat at the same time, so tonight Harry is having his dinner in the kitchen ... with the cats. And Mum is being the Queen, sitting at the head of the table and making sure everyboddedy behaves their-actual-selves. And doesn't bark. Or hiss. She's on a quite actual serious dipple-O-matic mission, and if that means she's got to pretend to be Queen Mum of the Kitchen, then that's wot she's going to do.

Harry did quite very well at concentrating on his dinner, and he wasn't bothered about eating in the same room as the cats. And cos he was actual quiet, it meant the cats could get a Good Look At Him, and Mum could get a Good Look at Them getting a Good Look at Him. As far as I are concerned that's far too many Good Looks to be healthy, so me and King Dad of the Office hidded in there until it was all over.

Gandhi and Mouse and Gipsy were not a-very-tall bothered by Harry once he was quiet. They did eatering their dinner, and then having a wash, and generally making it clear that so long as Harry didn't bark, he was welcome in their house. And then, when it was time for the cats to go outside, Queen Mum scattered a few treats on the floor so Harry was much, much more actual hinterested in finding the bits of grated cheese than chasing the cats as they slidded very, very quietly out of the cat flap.

Mabel did politely decline Queen Mum's kind offer to come to dinner, which was to be hexpected, but I was actual surprised about Frank. I always fort Frank was fearless General Frank of the Ginger Militia, and Defender of the Cat Flap, but it turns out that Frank is a bit of a, well, wussy-pussy. He is perfickly

112

continued page 121

Being next to the wall meant I did have to very get used to peoples clambering over my boat. Dad did exerlent helping me with this fing.

Night-time strolls and sniffs in Zeirekzee

On the way to Goes-goose-house-however-you-say-it.

My sun shelter in Zeirekzee.

Paddling at the Island.

Locks are fuge but I are now
hexperienced at them, and my
famberly is hexperts.

Hinternational dipple-O-matic
duties in Ooudorp with Pepper

Being a confident boykin doing exerlent Lookering. Out.

Relaxing in Goes-goose-house-however-you-say-it.

Sailing with my Dad.

Harry: my foster brother.

On the beach with Harry ... and wondering if he has any brakes ...

Being stucked in a beach hut is Not. Funny. A-very-tall.

Harry and Fran
smoochir

Harry showing
off his
footballing
skills.

Harry and his Dad-
wot-luffs-him.

The very broked banisters.

The previously ginger one with her Naward.

Getting hambushed by Harry on Southwold Beach.

Routine is more actual himportant than Crispmas Day, I do fink.

Me and Harry.

happy to defend our house from all the neighbourhood cats, wot are all smaller than actual him, but he is terry-fried of Harry. He sat on the window sill cowering, and tried very actual hard to melt into the window frame, until Queen Mum tooked pity on him and opened the window so he could run away without losing his digger-nitty.

Queen Mum of the Kitchen says that there will be another dipple-O-matic meal in the kitchen tomorrow morning. And everyboddedy will be behaving themselves again. There will be no barking and no hissing. And there is no need for Dad to curtsey ...

September 21
Harry, Mouse, Gandhi, and Gipsy has reached a Hunderstanding. They has all decided to do hignoring each other. Unless any of them do anyfing completely actual redickerless, like get too close, or fall off a work surface. All the barking has stopped and peace is halmost reigning.

Mum says she's quite actual glad about this. She's rubbish at all this cool, calm, collected stuff, and we've run out of crisps and ketchup. When she HAS to concentrate on pri-orry-trees like dipple-O-matic missions then she can, but she can't do anyfing else at the same time.

September 22
I fink we is going to need another House Rule. Dad reckons if peoples did following his completely-hignored-but-probababbly-quite-useful Rule about Keeping the Kitchen Door Shut, we wouldn't actual very need a Rule about No-Harry-on-the-Table.

September 23
The noo dog group is going to be called SARDOG, wot stands for Southwold and Reydon Dog Owner's Group. Dad and me fink it shoulda binned called Southwold and Reydon Doggies and Hoomans-wot-like-Doggies Group, but Mum says that's not as catchy, and dogs don't know how to vote, anyway. And it's quite actual himportant that the poxy Cow Sell take this group seriously. Dad reckons I'd be fabumazing at voting, and telling the peoples on the Cow Sell wot to do. If everyboddedy lived their life like Worzel, the world would be a happier place, wouldn't it Whizzy Wookins ...?

Mum finks Dad shouldn't have had that last bottle of cider.

September 24
Frank has decided, after a lot of fortful finking, that he is going to have to trust Harry. Either that, or he is going to have to stop rolling around on the kitchen table, wot is his very quite favouritist fing to do.

Today, Frank was quietly wriggling about in the sunshine when me and Harry strolled past on our way to the garden. Harry stopped when he sawed Frank, and I fort things were going to get all barky and hissy and hun-dipple-O-matic again.

I fink perhaps Harry mistooked Frank for somefing else, cos he didn't look much like a cat; more like a walrus hauling himself up an actual beach, so he had to give Frank a little sniff to work out hexactly wot had landed on the kitchen table. Once Harry had worked out that Frank wasn't hedible or a walrus, he carried on following me, even though Frank's extra long tail flicked him on the ear as he went past. Harry didn't react, and Frank was too busy barbequing himself in the sun to fully happreciate wot just happened, so I reckon the worst is over. With any luck, Harry and Frank might soon become friends.

September 25

Friends, I said. Just flipping friends was all we did ask for! But now Harry and Frank has decided not only to be friends, but to do kissing, little nose-touchy kisses, and Mum finks they is dorable, and also come-and-look-at-these-two, and bring-me-my-camera.

I hope she doesn't fink I are going to start kissing the cats. I'll give Gandhi a bit of a pokey tummy nudge every now and again when I want to have a playtime, or when he's lying where I want to be, but I aren't kissing no blinking cats, no matter how many bitsa cheese it seems to be worth.

September 26

There is a new Queen of the Kitchen, and she's even got servants, hooman servants who Queen Mabel has trained to come when she calls them. Or at least hoomans who she has trained to call her to come when her dinner is ready. Then she sits on the fence just outside the kitchen door, and will allow them to pick her up and hescort her into the kitchen to her special spot where she wants to eat her dinner. And then they is actual allowed to carry her out again, and put her back on the fence.

Queen Mabel has decided that she is not going to have nuffink to do with Harry a-very-tall. He is beneath her digger-nitty. And all the time she is getting carried from where she is to where she wants to be, Harry is about three feet beneath her. If she doesn't ever look down, she can pretend he does not hexist.

September 27

Last week, the previously ginger one did manage to get a bird to do a poo on her head. Happarently, this was Orrendous and Hembarrassing, and did require Mum to Do. Somefing. About. It. even though she was thirty miles away from the previously ginger one, and also not really actual quite sure wot that Somefing might be. Apart from pay for the previously ginger one to have a hexpensive trip to the hairdresser, wot Mum did not suggest, even as a joke, cos the previously ginger one would have said yes. Himmediately. In actual very fact, Mum was just very gob-smacked that the previously ginger one didn't fink of this cunning plan ...

Instead, Mum did hexplain to the previously ginger one that getting bird poo on your head was actual lucky, and put the phone down before she could do disagreeing with her about it. And I has got to say I do very hagree with Mum, and I are quite actual pleased about this, because it means that Mum is starting to hunderstand my forts and feelings, and Seeing The World Like Wot I do.

I like bird poo. I hespecially like fesant bird poo, and also Nowl Pellets, wot aren't really actual poo cos they come out the other end. But the end result is roughly very actual the same. Small, fabumazing smelly bits of stuff wot I can rub all over my beard and cheeks, and most actual himportantly of all, noboddedy can see it after I has dunned this fing.

For weeks now, Mum has been finking me and Kite has been having a silly playtime, and rolling about being cute and luffly together without being able to see nuffink in the grass to complain about. And please do believing me, she has looked. And looked ... But without hevidence of disgustering, good-grief-wot-is-that-all-over-your-face, there has not binned any reason for her to decide a bath is necessary. All in-very-all, it has binned a fabumazing summer.

Everyfing wented wrong a couple of days ago, though, when a fesant did a poo on the patio, and I did make the ignormous mistake of rollering in it. And instead of it breaking up in the grass, it did just stay on the patio for Mum to see. I did fink about destroying the hevidence by eatering it, but I wasn't fast a-very-nuff. So me and Kite has been rumbled, and our rollering around in the grass isn't considered cute any actual more: it's causing shrieking hystericals, and then hencouraging tricky bitsa cheese from Kite's mum whilst Mum hinvestigates and digs around in the grass ruining everyfing.

So, now I has tooked to lying around outside for hours, and has just got to hope a bird does poo on my head like one kindly did for the previously ginger one. Wot is most unfair, cos she did not do happreciating it a-very-tall. Like I woulda dunned, and considered myself very lucky hindeed!

Visit Hubble and Hattie on the web: www.hubbleandhattie.com
www.hubbleandhattie.blogspot.co.uk • Details of all books • Special offers
• Newsletter • New book news

OCTOBER

october 1

It's National Black Dog Day today. Happarently, dogs wot is black do find it hard to get dopped for some stoopid reason wot I don't actual hunderstand. I aren't black. But Harry is, and he is booful. And Gipsy-the-foster-fridge is black, and so was Kes ... Over the years, my famberly has dunned fostering lots of black dogs, and they has all binned actual fabumazing. So, if you is considering dopping a doggy, please don't overlook the black ones. Fanking oo kindly.

*****BLACK DOGS*****

- For some reason, black dogs do not get dopped as quickly as dogs wot are other colours
- Black dogs are hexactly the same as other dogs, hexcept they is black
- Some peoples call being dee-pressed 'the black dog.' They has hobviously not metted Lola. Or Gipsy. And hespecially not Harry
- Taking pictures of black dogs is quite actual hard. Noboddedy can call them-actual-selves a good photo-taker unless they can get a good one of a black dog
- Black dogs should not lie on stairs. Or outside bedroom doors, cos they will get trodded on
- On the other paw, black dogs can sneak onto beds in the middle of the night and get away with it quite very much easier than ginger ones. Wot isn't funny
- Black dogs do generally have black nails. Wot means that honly hexperienced peoples should attempt to cut them, and they does not get wobbly mums finking that they should be able to cope attempting to cut off their toes
- Black dogs do get just as muddy as ginger dogs, but it doesn't show as much, so they don't get tolded they need a bath
- Black dogs can hide in the shadows and do exerlent mole-hunting, whilst hoomans spend ages lookering for them. Ginger dogs get seen much much quickerer. And tolded to go inside
- Being a black dog I do fink must be actual quite betterer than being a fuge ginger hobvious doggy wot can't get away with very nuffink

october 2

You is never, hever going to actual believe this, but Harry very actual likes going in the car, even after he had to live in one for two months. Going in the car is not my most favouritist fing in the world, so I don't fink I'd ever, hever want to get in a car again if I'd had to put up with that.

Harry is quite very willing to go in the boot space, as well: he just hopped up into it like it was the most natural fing in the hentire world. Which it is NOT. Gettering into the car is somefing that has to be finked about carefully, and there has to be cushions and the hole back seat available before I will get in. And heven then, I get into the car like I has just actual hagreed to eat a lemon, Mum says. I tolly-rate it because usually we go to places I are quite actual keen to arrive at, like today, when we went to the beach.

I hope Mum doesn't start to fink that the way to get me to like the car is to live in it with me for eight weeks. She's rubbish about putting her sweet wrappers in the bin, and remembering to take out the coffee cups. There's barely any room in there at the moment, and we only spend about twenty minutes a day in it. After eight weeks, I don't fink we'd be able to see out of the windows.

october 3

Today, me and Harry has dunned Modder Nart, which mainly hinvolves actual ripping as many different fings into tiny ickle bits as possibibble, and leaving them scattered about in Nartistic patterns.

Our materials of choice has binned the hole of the recycling bin that Harry managed to open. Then, when we had dunned Modder Nart, we also did Per-foreman's Art, wot is making as many barky, yippy that's-another-very-one-deaded-forever noises as possibibble. Making Modder Nart, I has discovered, is not as easy as it looks, and I are quite actual hexhausted.

I would like to be having a snooze now, and hadmiring our Whirr-kof-Nart from the sofa, but that has not been possibibble, so I are sulking. Mum says our house isn't the Tate Gallery, and Dad will be home soon, and wondering where he can sit. Or stand. Or even if he wants to come home. She's got to do tidying up whether we like it or actual not.

Usually, I have a snooze whilst Mum does tidying up, but she has decided that the bestest tool for clearing up our Nart is the garden rake, wot I was not hexpectoring. Snoozing whilst someboddedy does flailing a garden rake around hindoors is himpossibibble. I does not like the garden rake when it is in the garden, and I are even less keen on it when it is in the house. It is scary and fuge, and Mum is actual useless at standing it up safely, so it falls over with a fuge blinking clang. Happarently, I are to be grateful that it's honly the garden rake she is using. She says a snow plough would have binned a betterer tool, but she hasn't got one. And I would have liked that heven less.

The hole hexperience has binned Orrendous, and I are finking seriously that I will retire from being a Modder Nartist. Harry isn't planning to retire: he's just waiting for Mum to clear some space so he can start again on a cleared carpet.

october 4

Two birds managed to fall down our chimberly today. At the same time. I fink Starlings must be like sheep, and blindly follow each actual other around, cos I find it quite actual hard to believe that two would hindependently fink going down our chimberly was a good hidea, not when there is five cats and me and Harry waiting for them to arrive. Either that or they was having a squabble wot went horribibbly wrong.

I are pleased to say I was very actual low down on the list of pri-orry-trees of who-to-get-out-of-the-room first. Harry was first on the list, mainly cos he would not stop barking and getting hunder Mum's feet, and was too actual keen to stick his nose within stabbing distance of the Starlings' beaks. Then the Starlings decided that their hargument wot they was having up on the roof needed to be postponed while they dealt with their common henemy, wot was Harry.

Harry's barking woked up the fuge ginger boyman, and he did decide that, despite the fact that he was half-asleep, and was halarmingly-and-very-quite-honly wearing a pair of boxer shorts Mum fort she'd chucked away about ten years ago, he could help by trying to catch the Starlings, half-naked, with his hair flopping about all over his face.

I are quite very sure that it is in the Rules of Being a Mum that you is NOT allowed to call your children half the words beginning with B that the fuge ginger boyman got called. And neither is you allowed to tell them to go away and Take Harry With You, using the Other Words Mum choosed, but he was being a blundering hidiot cos he hadn't tied up his hair. So he strode off in a sulk, muttering fings about just-trying-to-help, and me-and-Harry-will-be-in-the-kitchen.

We is all used in dealing with one Starling in our house – it happens quite very often – but two was a hole noo hexperience. Once Harry and the fuge ginger boyman had dunned bogging off, the Starlings remembered about the squabble they was having with each other on the roof, and decided to start attacking each other instead. Wot lead to a lot of bird poo all over the radiator, but it did mean that Mum was able to finally open the window. Then she used a hempty cardboard box as a riot shield, and shoved them out of it.

october 5

Mum has said sorry to the fuge ginger boyman about the somefing-beginning-with-B-words, and the Other Words she said to him. She has been being The Mum for wot seems like forever now. She was not much older than the fuge ginger boyman is now when she became The Mum, and she's had a-very-nuff of everyboddedy expecting her to be The Mum, the Hole. Time.

The fuge ginger boyman needs to remember she was a hooman before she becamed a Mum; the previously ginger one needs to try and get out of bed more often, and Dad needs to remember she is not just a provider of ketchup and cuppateas, SHE IS A WOMAN, too, and everyboddedy needs to stop taking her for granted.

And she's scared of Starlings.

october 7

I tooked Harry over to meet Kite and Maisie today. Maisie was not himpressed. A-very-tall. And Harry was quite actual wary of her. But for all her I-do-not-like-you-strange-black-dog barking, and stiff-boddedy-standing-and-staring, Mum and Kite's Mum did quickly realise that Maisie was the most frightened of all.

I did find this quite actual surprising. Harry isn't frightening: he's a daft, over-hexcitababble hidiot, and Maisie's reaction to him was not pro-porch-in-all. So Mum and Kite's Mum had to do a lot of exerlent hinterfering by putting themselves between Harry and Maisie. And stuffing them both full of treats.

Fortunately, Harry and Maisie is both quite very hinterested in treats, and they was so busy watching the treats, and then eating the treats and henjoying the attention, that they did soon forget they was scared of each other. And then Maisie went back to her usual sun spot under the table, and me and Harry did do a demonstration of Sighthound zoomies, wot is almost the same as Worzel and Kite zoomies, honly ... flipping Heckington Stanley ... that boy is fast! He is the fastest Sighthound I has ever, hever met, and I honly did winning cos I know my way round the 'salt course in Kite's garden betterer than Harry.

october 8
Mum's got a job! I aren't very allowed to say a Proper Job, cos when Dad tolded someboddedy on the phone tonight about Mum's Proper Job, he did get A Look, and Mum started to reel off a list of Not-Proper-Jobs she currently does, and wondering if Dad would like to do them instead.

The most himportant fing about Mum's job is that she can do it from home, so I aren't going to be actual habandoned to do lookering after Harry On. My. Own.

Mum's going to be writing for a maggy-zeen called *Dogs Monthly,* wot she is very, very quite actual hexcited about. Apparently, Mum used to do writing when she was at Universally, so she can just about remember wot words to use, and how to use them.

To celly-brate her noo job, Mum has had her nails painted strange colours. Dad says he can't see the connection between having painted nails and having a noo job, hespecially as she's going to be sitting at home, and noboddedy will see her nails, but that's not the point, Mum says.She is feeling quite actual proud and pleased with herself, and she wants to look as good as she feels, cos when you spend all day doing Not-Proper-Jobs, you can start to feel unhimportant, and not worth bothering with.

october 9
Maisie forgot that Harry wasn't frightening, so we had to go through the hole blinking Pavlova-with-the-treats-and-the-attention again tonight. But it did not take so very actual long this time. And Harry was less bouncy and redickerless, which made fings much actual easier. Mum reckons we'll have to keep doing this fing for a few days until Maisie manages to remember.

Or until Maisie starts to hexpect treats every time Harry does appear, wot won't be good for her diet.

october 11
Very actual sometimes, when you is training your doggy, it is worth calling in a hexpert. So today, I did take Harry to meet Doris, Kite's cat.

In our house, Harry has gotted the hole leave-the-cats-alone-don't-bark-and-Happroach-With-Caution way of finking boofully. And our cats have decided that he is as hunimportant as Worzel Wooface

But out of the house, our cats run away. Cos they can. Which makes sense, really, when you do fink about it. They tolly-rate Harry in the house cos that's where their food is, and where they want to sleep. But outside they can choose. And they choose not have to deal with Harry, wot I do fink is wise, and all about having an Easy Life.

Doris, on the other paw, is not hinterested in having an Easy Life. And isn't wise. She doesn't need to be. She is either hincredibibbly brave or a bit fick, or a comby-nation of the two, cos when she sees a luffly boykin coming towards her, she just sits there. And it does not matter how much barking or play-bowing or dancing you does do near Doris, she Does. Not. Move. Hunless you touch her. And then she over-reacts Big. Time. Badly. And it gets really, wheely painful.

After about five minutes of trying-to-get-Doris-to-play, and another five of pointless barking, with me watching from a very safe don't-say-I-didn't-warn-you distance, Harry has decided that cats is really, wheely not worth the actual hassle, and he do still have his hi-balls, wot I do fink is a miracle.

october 12

If you did have a person in your famberly who was seriously very actual unwell, I hexpect you would call a hambulance, and then the men-in-green-with-flashing-lights would do turning up, and would take your poorly sick hooman to an Opital where they would get the help they needed to get betterer.

When you has a Men-Tall Elf, though, you has to call several peoples to get somebodddy to eventually answer the phone. And then, you might get tolded to drive to an Opital thirty miles away. And then after a wait of an hour, a doctor will come out to see you, and say that the previously ginger one is very actual unwell, and needs to come into Opital. Hexcept they don't have any beds. In very actual fact, there is honly one bed in the hentire East Angular left, and if they do waiting for a hambulance to come and collect the previously ginger one, that bed will have goned.

The doctor is very, very, quite actual sorry about this, but the bestest fing for Mum to do for the previously ginger one is to drive to Kings Lynn, wot is two hours away, Right. Now. At 11pm. So that's wot she did. And that's why peoples in my famberly call the poxy Guv'Ment, the Poxy. Flipping. Guv'Ment. And it's also why I hasn't had my walk today.

october 13

Tonight, Kite did spend a lot of time staring at the ceiling of her summerhouse, wot is a small shed fing that everybodddy does hide in when it's summer, but doesn't feel like it. Or when it isn't summer, and everybodddy does wish it was. At first, nobodddy could work out wot she was so quite very hinterested in.

Heventually, when Kite could not make anybodddy hunderstand wot was so very himportant, she decided to take matters into her own paws, and tried to climb onto her mum's head to get to the ceiling. This was very not happreciated by her mum. Or by Kite's Dad, who ended up having to do leapering up to rescue the flying cuppatea.

After everybodddy had decided that Kite was being redickerless, and asked her very quite nicely to bog off and play with Worzel Wooface, the hoomans did work out it was quite very possibibble that Kite had suddenly noticed the little lanterns on the ceiling. If you is a Labrador and hobsessed with having fings chucked for you to collect, they could be mistook for tennis balls, dozens of tennis balls just stucked on the ceiling to confuddle a very hobsessed and really-needing-to-get-another-hobby, Kite.

october 14

Kite did her staring-at-the-ceiling fing again tonight. It was my himportant and hurgent work to try to make her forget about the tennis balls on the ceiling before any more cuppateas got spilled. I did my bestest bitey-facey, and Mum found a noo tuggy-toy for us to play with. Harry tried to help but he was too actual hinterested in wot Doris was doing hunder the shed.

Kite's Dad did try to distract her with lots of hobedience stuff like fetch and back and down and wait, and she was a Good Girl, and all sorts of other clever words like that. Then she went into the summerhouse, and sat gazing at the ceiling again so had to put up with being called a fick Labrador plonker dog as well.

I do fink it is most unfair that when Kite is clever it is cos she is a girl, but when she is stoopid, it is cos she's a dog. This doesn't seem to be quite actual reasonababble ... Mum says I'll have to live with it, just like half the hooman race has dunned for the past billion years.

october 15

In our house, luffly boykins are allowed to do sleeping on the big bed with Mum and Dad, under certain very actual conditions, wot are these –

****BIG BED CONDITIONS****

🛏 The big bed belongs to the hoomans. Luffly boykins can sleep on the bed, so-very-long as they remember this fing

🛏 If Mum wants to turn over or poke Dad to stop him snoring, luffly boykins should do moving himmediately so that she can actual do this

🛏 The cleaning of my gentleman bits is not allowed. Ever. Nor is lickering of feets or scratching of collars. These is private fings wot should be dunned elsewhere

At first, when Harry was learning to trust Mum and Dad, Harry did follow these actual rules perfickly. But recently he has binned struggling. For a very quite start, Harry Will. Not. Move. He acts like he's had a Nana's-fetick when he is asleep, and does not do leapering up like he's binned helectrocuted when he does feel a toe in his ribs.

Like wot I do. Wot is exerlent, according to Mum.

He does also fink the bed is where he should do his washing, which sounds like a couple of teenagers snogging when he does that fing, but is not a sound that anyboddedy does want to hear that often. So he has binned hejected from the bedroom, and from tonight he is going to have to do all his sleeping downstairs.

october 16

This morning I has got a fuge Complaint to the Management.

I does not know if you has hearded the sound of a Saluki-Lurcher wot is Not. Getting. His. Own. Way, but it is a high-pitched, supersonic squeak, like a rusty bike. Harry has spended all night whining and whinging and whimpering, and all sorts of other words beginning with WH. 'W-h-ill you shut up, Harry' does not begin with WH unless it is said by a quite actual fed up and knackered Mum, and then it really, wheely does.

Wot with the squeaking and Dad's snoring, and Mum not being able to actual cope ANY-somefing-beginning-with-B-MORE, tonight, it has becomed my himportant work to Keep Harry Comp-knee. On the wrong side of the door. So everyboddedy can get some sleep. Wot I aren't that himpressed about. So I are sulking a bit … a very lot, to be actual honest.

october 17

Harry has decided that the WH words aren't getting him very far, so his current lemme-back-on-the-big-bed-plan is to stick his nose down by the gap at the bottom of the door and HUFF. And PUFF. And generally try to blow the door down. He finks it works cos after ten minutes of huffing and puffing, Mum generally gives up and opens the door. But all she does then is take him downstairs to his special armchair and tell him to go to sleep.

I would like it to be known that I has binned being perfick. I has got sulking-in-my-bed and hignoring Mum every time she comes down to park Harry on his armchair off perfickly. I has also added lookering through my hibrows with a hurt and disappointed face, just in case she is missing the point that it is REALLY, WHEELY HUNFAIR that I has got to sleep downstairs when I has dunned NUFFINK WRONG.

Dad is hobviously on MY side. He's very quite deaf in one of his hearholes, so he can lie on the other side, and not hear all the whining and whinging, and the huffing and puffing. But he can't turn off Mum wafting the duvet, and letting a cold blast of air hunder the covers every ten minutes, and tonight he did shoutering at her for waking him up. Again.

Dad never, hever shouts. So now Mum's knackered, feeling guilty and very, very hupset. And also sleeping on the sofa. With me.

october 19

Southwold and Reydon Dog Owner's Group (or SARDOG, as we is all calling it) is Hofficial! There was a Not Secret meeting tonight, and lots of people camed to hear wot was planned, and how the Cow Sell was being quite actual helpful now that it has a proper group to do consultering with.

Polly-Tricks is beyond me, I has decided. As far as I can actual work out, the hexact same peoples who hobjected to the beach ban and gotted hignored has dunned calling themselves somefing wot sounds like a superhero, and now, all-of-a-very-sudden, they is being listened to.

Auntie Charlotte is very relieved cos if fings carry on the way they is currently going, she she won't have to stand in an Helection, and there is a good chance that SARDOG will be able to save the beach, at least during the colder months of the year. But it has all got to go to a vote at the Cow Sell, so noboddedy should start celly-brating just yet.

october 21

As part of the Cow Sell consultation, Auntie Charlotte says everyboddedy who is a member of SARDOG has to do writing to hobject to the Beach Ban.

Cos Mum is hofficially the SARDOG member in our house, and cos Dad's spellering is hawful, we has decided to let Mum do the letter writing. This is wot she has said –

Dear Sir or Madam

I am writing to object to the proposals with regard to Southwold Beach. Perhaps you are unaware of the community of elderly people who, with their dogs, meet every morning to walk along the promenade? Joining them are often people with buggies and wheelchairs. This is a REAL community - a group of people who clean up after their dogs, catch up/keep an eye out for those less mobile or elderly, and use the beach as a public open space: something that the council is well below the national guidelines in providing.

It is unrealistic to imagine that such a 'ban' will be patrolled or that fines will be issued. After all, if the council had been capable of achieving that, perhaps people would be fined for the existing dog fouling issues. All that will happen is that those who currently pick up after their dogs will obey the ban, and those who are either unaware, such as tourists, or non-compliant about picking up, will continue their existing practice without the eyes of more community-minded folk to shame them into picking up.

The proposals are unnecessary, unenforceable, and ultimately will damage the local economy. Families and people with dogs will simply choose to go somewhere else where they can park their car, and walk their dogs easily and without fear of fines.

I urge you to reconsider this ban, and keep Southwold dog-friendly.

Kind regards

Worzel's Mum

Honly, she didn't sign it Worzel's Mum. I changed that bit so you didn't get all confuddled and fink I'd borrowed someboddedy posh and clever who knows lots of big words, HAND knows how to spell them, to write the letter. And then pretended that she was my Mum.

october 24

I cannot do recommendering hearhole plugs a-very-nuff. Or stair gates. I should probababbly not be recommendering these fings that have helped to solve the Harry problem, wot in turn means that I are stucked downstairs for-actual-ever, but sometimes a luffly boykin has to cut his losses and decide to make the very actual best of fings. And the actual best of fings currently is Harry learning how to go to sleep downstairs, and not huffing and puffing and whinging and whining. Cos it was getting very actual catchering, and all my famberly were starting to huff and puff and whinge and whine cos they weren't getting a-very-nuff sleep.

And neither was Harry. Doggies need to sleep as well: proper good sleep so they is happy and ready to face the noo day, running on the beach and all the other hexciting fings that could be planned if everyboddedy wasn't so blinking knackered.

Like everyfing in our house, getting the stair gate up was not simple. Ever since Gipsy-the-foster-fridge bashed into our banisters, they has binned wobbly at the bottom, so the stair gate has had to be put at the top of the stairs. Mum did point out that if Dad actual Fixed. The. Banisters like she's been asking him to do for the past two years, then not honly could we have the stair gate at the bottom of the stairs like any normal famberly, but he wouldn't have had to come up with some bonkers contraption at the top of the stairs. And it probababbly would have tooked a lot less time.

I fink Dad has got almost as much stubborn Saluki in him as Harry.

october 26

Does you know the word Hindsight? Hindsight is when you realise that you should have putted Harry in a crate to sleep from the first night he arrived. And not fort like some soppy hidiot that because he'd been sleeping in a car for two months that he would prefer to be a bit more stretched out at night. And also completely missed the blinking point that Harry HAD actual gone to sleep in the car, and not paced backwards and forwards wondering wot he was missing out on hupstairs.

But it's all too late now. For a very actual start, Mum's lended out all of her crates, but can't remember who to, and the honly reason she has been able to fink about these fings is cos Harry finally managed to stay downstairs on his armchair all night without any whinging or whining, and she's had a proper night's sleep.

october 27

Me and Harry has binned naughty boykins at Kite's house.

I aren't naughty very actual often, but sometimes I do seem to have a Moment Of Madness where I do do lots of naughty fings all in a row. And I did do that fing yesterday.

First of-very-all I did find a plant pot, which I did take for a trip around the garden. And Harry did have a go after I showed him how much actual fun it was. Hunfortunately, the pot did still have a plant in it. Wot was special and himportant, and very quite hembarrassingly Not. Mum's, but Kite's Mum's.

Kite's Mum did do running-like-a-penguin after us to try and get us to do giving it back, wot we did heventually, but not until Mum had dunned quite a bit of laughing at Kite's Mum and her rubbish running, and not a lot of helping get the pot.

Then I did find a car-washing sponge. We does not have car-washing sponges cos we does not have a posh car, and Dad's always fixing the boat, anyway, and Mum says she isn't washing no cars cos that's wot rain and Dad's-in-the-Doghouse are for.

Did you know that car-washing sponges do falling to very bits in your mouth? Hespecially if you does give half of it to Harry to have a bit of a tug about with.

It's all Maisie's fault, I do fink. She does spend all her time rootering around in the plant pots looking for himportant fings wot might be nice to eat, but probababbly aren't, and she did give me the hidea. When Kite's Mum wented to put her plant-in-a-pot back on the shelf where we did nick it from, she did discover I had dunned careful selectering and rummaging of wot pot I did want, and had managed to knock quite a few of the others onto the floor.

So we is in the actual Dog-House, and Mum has binned wondering if she has any plants to replace the one that did get dragged around the garden, and is now ruined. Cos she knows she doesn't have a car-washing sponge.

But Kite's Mum's got billions of them now. Lots of little ones. All over her lawn ...

october 28

If there is a Dad in your famberly it is very actual himportant to make sure he hunderstands the Himportance of Being Consistent. Like, not letting Harry sleep on the big bed cos Mum isn't there. And finking you'll be able to get away with it, and Mum. Won't. Know. cos Mum. Wasn't. There.

Mum will know, cos Harry will spend the next two nights whining about Dad-let-me-sleep-on-the-big-bed, and whinging cos he's got to stay downstairs again now that Mum is back.

october 30

I'm feeling sad today. My very beautiful and growed up senior lady Friend in the North, Kelpie, has goned to the Rainbow Bridge.

When I did meet Kelpie, there was a small moment of wondering if she would do tolly-rating me, but Kelpie did very happrove of me a lot and a lot. Which was a good fing cos I didn't fancy sleepering in the car.

Kelpie was booful and graceful, and she did die doing the fing she loved most in the world, running around her favourite park trying to find a squiggle. She was 16 years old, and although there is never a good way to lose your bestest companion, I do hope that the fact she was doing somefing she loved will eventually help Our Friends in the North do smiling again.

NOVEMBER

November 1
****FUGE PROUD NOOS****

The previously ginger one has binned nommy-nated for a Naward. She has binned nommy-nated to win a Naward for Houtstanding Bravery, cos of her Fight the Stigma campaign. She says she doesn't hunderstand cos she hasn't dunned rescuing peoples from a building wot is about to clapse, and she doesn't feel very brave, just a bit sick cos she tooked her medicines at the wrong time of day, and she needs to finish her collie-um-num before she can have a lie-down.

Mum's tried to hexplain that you can be brave-in-the-head as well as brave-in-the-boddedy, but all the previously ginger one wants to know is if she can have a cuppatea, and wot should she wear cos the Naward Ceremony is honly a month away, and she needs to start panicking about it. Right. Now.

November 2

Dear Auntie Stella

Fank oo very actual quite kindly for the ball wot you did send to me.

Me and Harry did take the ball round to visit Kite, and she and I did fink it was fabumazing. We did playing with it in the dark so there is no pictures, but it did make Kite's face go all sorts of funny colours when she was carrying it. She did look like a doggy disco, wot did make me fink she was not Kite for a bit but I did get actual over it, heventually.

Today, I was lookering forward to seeing Kite and her disco face, and playing with the ball ... but I can't. Cos Doris has nicked it. I did not know that cats like balls. None of the cats in this house like to play with balls. And now we has gotted a fuge problem cos she won't give it back, and I do fink that is most very quite unfair. It's MY ball. But Doris is a cat, and I aren't going to try to get it off her cos she will turn into a psycho-ninja-hissing-fing-with-knobs-on.

Kite reckons she'll be able to get it back heventually, but I are being a wise Worzel. And staying right very out of it.

From your luffly boykin

Worzel Wooface

November 3

I love Mud. Mud is wot you get when you mix rain and me. I seem to be a mud magnet. I can collect it wherever I go. And it sticks to me much betterer than it sticks to hoomans. Even when me and Mum go on the same walk on the same bit of countryside, I come back covered in mud, and Mum doesn't. Unless, of course, I do forgetting to look where I'm going, and barge into Mum at 40 miles an hour.

Then she makes mud almost as well as I do. And a lot more squeals.

All the rain we've been having recently has made it very much easier for me to collect mud. Mum says that some dogs are trained to stand on a towel and have their feets wiped before they is allowed into the house. I are not one of those dogs. I are one of those dogs wot is trained to make a swift and cunning sprint up the stairs, and onto Mum's bed. I generally use the time Mum is trying to get her wellies of her feets to make my move. She is very rubbish at running up the stairs with one wellie off, and the other half-on and half-off her foot, flapping about like a dead fish. Then all she can do is hope someboddedy else is upstairs when she frantically yells make-sure-my-door-is shut, and don't-let-that-dog-on-my-blinking-bed-again.

I fink this is very quite actual unfair. Beds is perfick for getting mud off me. I don't know why people's bother with the towel fing. Duvets are much, much betterer, and it doesn't leave *that* much mess. It really doesn't show up nearly as much on the duvet as it does on me.

Apparently, this is all hunacceptababble. I are not allowed to go upstairs with my muddy magnet body anymore, and all the bedroom doors are being shut like some military hoperation before we even leave the house. And Mum is trying to get me to do standing on a towel in the kitchen to get rid of the worst of the mud before I spread it everywhere, but it's not going well. The towel is rubbish.

But it's all very okay ... turns out the sofa is almost as good as the bed ...

November 4

It may be November, it may be a very bit cold and windy, and a lot and a lot rainy, but I are not ready for my paddling pool to be putted away. Apparently, this means I are bonkers-crazy.

Harry finks the hole concept of paddling pools is bonkers-crazy. In very fact, Harry finks water is for drinking, and that is-very-all. He will not go in the sea or in puddles, and he does look most actual hoffended if I do splash him. Even when it was warmer he fort this fing, so I has got no hope now it is getting actual chillier.

November 6

Kite's Mum says it's no actual good: my paddling pool has got to go away until next spring, and wot is actual worse, there can be no more zoomies in her garden at the very moment.

Me and Harry has completely and hutterly ruined her lawn doing our races, and she is actual quite sorry, but if we keep doing it, she isn't going to have any lawn lefted.

The problem is that me and Harry don't run in random places. Ever since he did arrive, and we did work out that we is halmost hexactly the same speed, we have been going round and round and ROUND the same route in Kite's Mum's garden, and there is no grass a-very-tall lefted in those places.

Mum says I are not allowed to sulk about this fing. We has got all sorts of places we can run and play, and fank goodness we is still currently allowed to go on the beach. Otherwise, Harry would go bonkers. He needs his runs and his zoomies every day whilst he is a redickerless teenager, so that is where we will be going.

But Harry says he is Not Going In the Sea. Not even up to his tippytoes. Mum finks this is very okay – it's one less hutterly soggy dog to deal with.

November 7

Dad's just phoned Mum from work. Dad never phones Mum from work unless it is really, wheely himportant, or unless his bottom does phoning Mum all by itself. Then Mum has to listen to Dad arguing with a lump of rope wot isn't doing wot it is actual told.

Anyway, once Mum had worked out it *was* Dad and not his bottom phoning, the hole himportant reason for the phone call did become clear. Dad did wonder if Mum had got a copy of the local noospaper? He said he hadn't had a chance to read all the words yet, but there was an ignormous picture of Mum and the previously ginger one Right. On. The. Front. Page of the noospaper, with the word

IMMORAL

plastered in fuge black writing hunderneath. And just hexactly wot have they done now?

November 8

Fankfully, Mum and the previously ginger one isn't IMMORAL, and everyfing is quite very actual okay. Dad says everyfing isn't okay: everyboddedy at work finks Mum and the previously ginger one has dunned somefing naughty, and so will the Hentire. World, unless they read the article properly. And will Mum please stop giggling and looking so pleased with herself!

Mum finks it's fabumazing. She doesn't care who finks she is IMMORAL if it gets people to talk about how wrong it is to send children fundreds of miles from home when their Ment-all-elf is making them very poorly sick. That's IMMORAL, and it's about blinking time the Hentire World knew it.

November 9
****HIMPORTANT SIGH-AND-TRIFFIC DISCOVERY****

Every six months or so, Mum goes to the Hair Choppers, and they dig about a bit and find Mum's face again, and make her look respectababble. Apparently, the hair chopper loves it when Mum comes into the shop, cos Mum doesn't much care wot the hair chopper does, which makes up for the fact that she turns up in her wellies, and refuses to do talking about whether she's going on holibobs, and yells at the hair chopper if she pulls Mum's hair. And generally behaves like a six-year-old.

The lady wot does Mum's hair is called Angel. And she must be to put up with Mum, to be quite actual honest. According to Angel, most of her customers love having a couple of hours being pampered and relaxing. Mum finks it is a waste of valuable gardening time; she is Not. One. Of. Them. People.

November 10

It is my himportant work to notice when Mum has been to the Hair Choppers. I have to say I are rubbish at that fing, though. She still smells like Mum, and makes the same noises as Mum. And apart from being able to find her face a bit easier for luffly kisses, she's still Mum. And there is no need for any fussing or bravely running away.

So I do fink that if your doggy does barking or hiding from you when you has binned at the hair choppers, it can't be because you look different. It could be cos you go to the hair choppers too very actual often, and do skip Himportant and Hessential walks, and your doggy is sulking.

Or it could be that you does really not like wot the hair chopper has dunned, and you is secretly giving off cross and upset hatmosfears.

Or it could be wot happened this morning. When Mum woked up, she forgotted she'd had her hair chopped. She did nearly fall down the stairs when she saw someboddedy strange looking at her in the mirror, finking we was being robber-dobbed. Then she tooked a fuge deep breath to do a bit of psycho-ninja-yelling-get-out-of-my-house ... when she remembered.

Maybe you do that fing. That I *did* notice.

November 11

Happarently, there is hair choppers for dogs wot is called Groomers, and they Do. Fings. To. Dogs. And their Mums and Dads let them. I are quite very sure this should not be allowed.

I fink I could do coping with their brushes and their combs. I has to do putting up with a bit of brushing at home when Mum starts whiffling on about me needing to Look Smart. I can't fink why I do need to actual Look Smart, though, it does not make me cleverer or faster, or help me do better snoozing. And most of all, it does not make me happy.

After about five minutes of being made to Look Smart, I do run away bravely and hide, and then it doesn't matter if I look smart or scruffy, cos noboddedy can see me under the table in the hoffice. Which is where I do stay until Mum puts the Looking Smart kit away.

Groomers do Other. Fings as well, like nail-cutting. Nail-cutting is probababbly hessential, and to be quite actual Put. Up. With. cos long nails can make a doggy's feets hurt, and stop him running fast. But if it is all the same to actual you, I'd rather Sally-the-Vet cut my nails. She spended seven years at Universally finding out where my nails stop and my toes start, and I do fink it is a job wot you can never, HEVER be too over-qually-fried for.

Most of wot Groomers do seems to hinvolve baths, and I don't want one of them, neither, fanking oo kindly. If I get sticky or very actual muddy, I are more than actual capababble of dealing with it. And I aren't being dried. I are never, hever being dried. I was borned with this shaking talent, and can wibble-wobble my skin all the way from my nose to my tail without any actual help. Groomers have these machines wot blow warm air about. I has had honly one experience with warm air when Mum opened the fan oven door when I was in the kitchen, and it did take me a week to agree to walk past the cooker again. It was Orrendous.

Mum reckons I have got it all wrong about Groomers. They is very well trained and actual quite kind, and some dogs do heven love going to be pampered and brushed and trimmed. But if such a doggy do hexist, I has not met them yet. There might be some dogs wot do Putting Up with It, or who are hogpigs and will do anyfing for a bitta cheese, but I are Not. One. Of. Them. Dogs.

November 12
Dear Barry-the-Cat-from-up-the-lane
You might want to fink about somewhere else to sulk while your hoomans are in France. Cos after last night's harguments, the hole-in-the-shed-door is being guarded by General Frank of the Ginger Militia. And now Mabel is too scared to go in there, and has been forced to take up residence in the room with the washering machine. It's all getting a bit musical cats round here, and that is Never. A. Good. Fing. Fanking oo kindly for taking my wise advices so we can get some sleep tonight.
From your luffly boykin Worzel Wooface
Pee-Ess: I has not currently binned asked to rejoin the Ginger Militia, which is a good job cos I aren't keen on the shed since Dad used the sander when I was not hexpectering to have to be brave.

November 15
I does not know why Mum finks I has binned drinking disgustering water from a bucket that has got God Knows Wot in it. I do fink it is very unfair for Mum to do casting nasturtiums in this actual way. I was quite very careful to make sure I was not hobserved doing this fing. I did exerlent sneaking off when Mum was yackering on the phone. I are innocent, and none of those gunky, smelly, no-I-don't-want-a-kiss-you-revolting-rummager fings. I are booful and perfick, and other words like actual that.

And I does definitely not want my beard fiddled with. Or washed. And I aren't going to one of them Groomers.

November 17

A MAN CAME TO SEE ME YESTERDAY

A man came to visit yesterday
He even brought some cheese
But I'm wheely actual sorry to say
I wasn't all that pleased.

Mum was quite hembarrassed
The Man had come a long way
She felt quite very harassed
And didn't know wot to say

The man was very gentle and kind
I liked him a lot and a lot
But his cheese I did quite actual decline
Mum fort I'd lost the plot

But then the Man did say the cheese
Was diet cheese and fantastic
So, nice slim man, I hope you agrees
That cheese it tastes like plastic

November 18
Has you ever tried to catch a frog? I wouldn't do wot you'd call recommendering it. It seems to be hot and tiring and actual quite pointless work.

The trouble is ... a couple of frogs are stalking Mum.

First of-very-all, they did suddenly jump up and shock the living wotsits out of her when she was diggering up some stinging nettles. When you is diggering up stinging nettles, it is quite actual hessential that you do know where all your arms and legs very actual are, so they don't do getting stinged. Having two frogs suddenly try to jump into your wellies does not make that an easy fing to do, and it ended about as painfully as you can himagine it might.

Anyway, once Mum had dunned getting stinged cos she flapped her hands about, and after she'd dunned giving herself a bit of a talking to about it being reasonababble to find frogs in the garden, and no, they wasn't punishing her for diggering up their home, and yes, they would survive just fine without the stinging nettles, Mum did her bestest best to forget about them.

Wot is quite actual difficult when your hands is having a hexploding stinging nettle itch-fest to remind you.

Somehow, and we isn't quite sure how, the frogs have now tooked up residence in the compost bin. Which is a quite very exerlent place to live if you does like slugs and snails, but a bit wot you'd call life-limiting for a frog, cos there isn't much in the way of water in there. And it must be pretty actual dark and smelly. And hot.

Mum doesn't fink this is a suitable henviroment for them, and has been trying to Rescue. Them. Trouble is, the frogs has made it stonkingly clear they does Not Want To Be Rescued. Every time Mum tries to catch them, they scurry to the bottom of the bin, but not out of the hole that Mum has lefted for them. And she's starting to take it a bit, wot hoomans call, personally.

Dad says Mum is being daft, and leave the poor frogs alone. Otherwise, when she gets stucked in the compost bin with her legs flapping about and her head in the stinging nettles, he's going to leave her there. Or take photos. He can't decide which, but would she please Stop. Trying. To Rescue the blinking frogs.

November 19

Harry's gotta talent. I used to fink I was very actual okay at football, but Harry is much, much betterer than me. Dad reckons he's betterer than most of the Ingerland team.

Wot isn't saying much, apparently ...

November 21

Yesterday, me and Harry did get very hinvited for a play date at Southwold beach. We did go to meet Mabel, who is a very actual fluffy Collie. I did fink Mabel was dorable, and she did like me very quite actual a lot and a lot as well.

Mabel did fink Harry was too fusey-tastic, and did tell him to bog off with knobs on.

Wot he did do heventually.

After we had dunned a lot and a lot of running on the beach, and Mum had dunned hexplaining that, yes, I was a quite actual long way away, but I would be back in about five seconds, and, yes, I can run quite ever so very fast, and after Harry had dunned not looking where he was going, and I nearly ended up with a large bitta Harry stucked up my bum, For Hever, Mum was asked if she did want a cuppatea.

In very general, I do approve of cuppateas. Cuppateas quite very often mean we will be staying out longer, and sometimes there is a treat for Worzel Wooface, so they is usually exerlent fings. How-very-ever, I did not happrove of this cuppatea a-very-tall. Cos I got Putted. In. A. Stable. With. Harry.

Peoples who do know me quite actual well, will know that when I was ickle, I did get called a Donkey and a Carthorse, and even a Camel once. I fort all this horse stuff had finished. Mum says it wasn't a stable, it was a beach hut, and I did have to stay in there while she sat outside for her cuppatea, cos Harry kepted on being a plonker about having his lead put on, and Mum wanted to drink her tea without chasing up and down the prom-and-hard saying Sorry-about-the-hidiot-Lurcher.

Which is all very actual well, but as soon as Mum did half turn her back, blinking Harry did do leapering over the stable door, and went for an epic round-around-and-fandango whilst Worzel Wooface stayed stucked. Cos I could not do jumpering over the door. I was not himpressed.

But wot Harry does not know is that whilst he was being a plonker on the beach, and Mum was doing gettering him back on a lead and bringing him back to the stable fing, I did get lots of treats and fuss, and tolded I was a very good boykin. Still doesn't make up for the stable fing, though ...

LURCHER BRAKES

I don't know who hinvented Lurchers
But they did make a few mistakes
Most of me is quite perfick
But I don't seem to have any brakes

Dad says it's a real design fault
A dog that's as fast as a car
Should have a better way of stopping
Before they've gone too far

Mum finks we've been quite actual lucky
So far, no-one has broken their knee
When my no-brakes are working as well as they can
And I'm flying towards them, carefree

My no-brakes don't usually bother me
It's the hoomans what hobject a lot
But today I do fink they may have a point
I've got to work out how to stop

Cos Harry has founded a new game
When I'm rushing along by the sea
He lies in wait, till I'm flying so fast
Then leaps up and hambushes me!

So if you do meet the hinventor
Of Worzels and Lurchers like me
Could you tell them I do need some brakes after all?
Cos nuffink will stop blinking Harry!

November 22
Dear Harry

Please do not poo on my Not-Grass-Really-Carpet. You has got plenty of other places to choose in the garden, and now Mum finks you've proved that it IS grass and not carpet. I do not fink you has proved that fing a-very-tall. You has proved that you is A Nidle Wotsit who can't actual be bothered to walk more than six feet from the back door.

And now I don't feel like lying on my carpet, heven though Mum has cleared it up.

From your luffly boykin
Worzel Wooface

November 23

Mum has decided to skirt around the hole pooing on the Not-Grass-Really-Carpet fing, wot I do fink is most actual hunfair. That's cos Harry is a good hinfluence, she says. It started off with the car. It is very quite actual difficult to be hupset about going in the car when Harry is so fusey-tastic, and hurry-up-Worzel-we're-going-to-the-beach, so I has binned finding myself strapped in and ready to go before I has had a chance to even fink about worrying about it.

And now, Mum reckons, Harry is being a good hinfluence about Hencounters with Hobjects.

There is lots of Hobjects I find a bit scary wot, according to Mum, I really, wheely should not. Like bits of branch that need pushing past or gates that look different.

Mum says I are too cautious for my own blinking good sometimes, and I need to worry less; if she is going somewhere or walking past somefing, then that Hobject is very okay, cos Mum isn't going to put herself in actual danger, and she wouldn't do that fing to me, neither. I fink too much, according to Dad. I need to fink less and henjoy more. But stopping finking is very actual hard, I do fink, hespecially when all that finking comes from somewhere deep hinside you from when you was a tiny puppy.

Harry doesn't find nuffink very much worrying, and apart from the hole sleepering in the car for eight weeks, his hupbringing has binned pretty much perfick. So when Harry sees somefing noo, he wants to find out about it. On the other paw, when I see somefing noo, I still generally would rather run away from it. And wonder if there is a different way we can go.

With peoples and other dogs, I are generally very okay nowadays, so long as they is gentle with me and do not try to be too huggy or fusey-tastic when we is first making friends. But Hobjects frighten me. And Mum can never predict what Hobjects I will hobject to, apart from sticks and wheelchairs and flip flops and flappy bags and humbrellas and signposts wot weren't there yesterday ...

But out on our walks, Harry's confidence is starting to rub off on actual me, wot Mum is quietly very actual pleased about. And it has got nuffink to do with being ripped in half by a Harry who wants to go in one direction and a Worzel who wants to run away in the other direction ...

November 24

Mum had to be a proper growed up hadult dog owner last night. In very general, I would rather she does havoiding this hole hadulting fing, but happarently it was for my own good, and if she could be brave, then I had to be as-very-well.

Mum. Cut. My. Nails. She says she's been trying to find the courage to do it for the past couple of weeks, but *she* hates it and *I* hate it. I have to sulk as soon as it is actual dunned, and Mum has to have wine. Not before it's dunned, hobviously, cos that might mean she missed or wented too far.

Hanyway, I has binned sliding about on the woody floor too actual much, and today when Harry and I was having a game of rufty-tutfy bitey-facey, I caughted him with my nail and There. Was. Blood.

Harry didn't notice the blood but he did notice the hurt, and he did fuge, loud, you-clumsy-oaf-Worzel-Wooface barkering up my nose. I did exerlent not-making-fings-worse by lookering away and not barkering back, so Harry did know I wanted to be friends again, but Mum said a nuff was a-very-nuff, and when fings were all back to peaceful and snoozy, she got out the nail clipperers ...

I does not know why I did tolly-rate the nail clippering last night but I fink it had somefing to do with Mum being determined, and finking more about me not hurting Harry, than worrying about chopping off my toe.

So I do fink Harry has dunned being a good hinfluence again but he isn't himpressed. Cos now Mum says she's got to do *his* nails, and he isn't having none of it ...

November 25

It's Black Friday. According to Dad, that's when everyboddedy goes bonkers when they remember it is honly a month until Crispmas, and they has dunned nuffink to be horganised.

Dad's got everyfing horganised. He horganised it years ago when he did marrying Mum. Job dunned, he reckons.

November 26

There is actual one Hobject Harry does not like. Dad finks it is the hexception wot proves the rule. In my hopinion, I fink Harry should just get actual over it. Or leave the room like wot I do. But no, Harry being Harry shouts about his hobjection, really, wheely loud. Well, he has to, otherwise noboddedy would hear him.

Harry is scared of the Hoover. Mum has been finking some probababbly not true forts that Harry's Dad did not do Hoovering, which is why he is so frightened of the noisy, sucky machine. Dad tried to tell Mum to stop being rude about Harry's-Dad-wot-luffs-him's cleaning nabbits, but as far as I know, my Dad has honly used the Hoover twice in his actual life, and when Mum sawed him doing it, she did nearly pass out with shock.

So maybe Dads using Hoovers is frightening, after-very-all. And the reason why they don't actual do it.

THE HOOVER

Harry doesn't like the Hoover
Doesn't like its blinking racket
Every time Mum does manoeuvre
it, he barks and 'tacks it.

Harry finks the Hoover's scary
And jumps up on the bed to stalk it

Watching it, he's very wary
He simply can't hignore it

Wish he'd learn to come downstairs
Like I do when the Hoover's roaring
Leave Mum sucking dust and hairs
His woofering's getting boring.

November 27

Dad has hurted his toe at work and he finks it is broked. Tonight, there has binned a fuge hargument about whether Dad should get his toe looked at by a doctor, and whether he can do anyfing about it, even if it is broked.

It's all the fuge ginger boyman's fault, happarently, though I has got to say that I are finding this hard to actual believe cos he is 200 miles away at Universally, and doesn't even know that Dad has hurted his toe. Mum says it all goes back to when the fuge ginger boyman was nine-years-old, and she's not-going-through THAT. AGAIN.

THAT. AGAIN is the bit that is all the fuge ginger boyman's fault, cos when he was an ickle fuge ginger boyman he did tell Mum that he had hurted his foot, and that he fort he might have broked it. But he walked up the stairs and across the landing to tell Mum this fing without limping or crying, happarently, so Mum told him to just run cold water on it, and they-don't-do-anything-for-broked-toes-you'll-have-to-put-up-with-it.

So, the fuge ginger boyman – who was a very good ickle boy – did wot Mum tolded him with the cold water, and did his bestest best to put up with it.

Three days later, he did decide to show Mum his foot, not cos he couldn't put up with it any longerer, but because he fort the colours were pretty cool. Which, if you do fink black with little dashes of purple and a hint of yellow covering your hentire foot is pretty, and not Mortifying and Hevidence of being a Bad Parent, is probababbly true.

But definitely quite very actual not cool.

November 28

I are actual quite delighted to hannounce that the previously ginger one is ... Previously Ginger again! Mum is *not* delighted, though, as her booful daughter's hair is currently a sort of dodgy pale yellow colour, with bits of green in it. It looks HAWFUL, Mum reckons, but the previously ginger one says Mum is not to actual worry: it is a work-in-progress and she'll love it when she's finished.

Mum is finding this very actual hard to believe.

November 29

Mum's cunning plan to stop Harry barkering at the Hoover is to scatter treats behind her when the Hoover is running. So far it is going quite very well, and Harry has decided that pouncing and woofering, and trying to actual attack the Hoover is not so much fun as rootling around for dropped bits of cat kibble.

I has decided that lying on the bed, very actual hignoring everyfing wot is going on with treats and peoples wrapped up in helectric wires, and trying-not-to-suck-up the treats with the Hoover, is a wise move.

And waiting for Mum to go into the next room so I can quietly finish up all the kibble bits Harry and the Hoover missed, is heven wiserer.

November 30

The work-in-progress wot is the previously ginger one's head, is, erm, progressing.

Sort of.

It is now halmost completely white, apart from a few very stubborn bits of ginger wot are refusing to be previous. She reckons she needs to give it one more go and Mum is to trust her. And no, her hair isn't going to fall out.

Mum is finding that very quite actual hard to believe, as well.

Visit Hubble and Hattie on the web: www.hubbleandhattie.com
www.hubbleandhattie.blogspot.co.uk • Details of all books • Special offers
• Newsletter • New book news

DECEMBER

December 1

Harry has stopped barkering at and attacking the Hoover, and Mum's cunning plan has halmost binned a marvellous success. Harry has now taken to doing a exerlent sit-with-a-bonus-waving-paw whenever the Hoover comes out. And then barkering like mad at Mum and the Hoover if the treats don't spew out fast a-very-nuff.

Mum's almost sure this *isn't* progress.

December 2

The previously ginger one has finished doing stuff to her hair and it is ... pink. Dad, being very well trained and after an easy life, knew hexactly wot Mum actual hexpected him to say about the pink hair, and has hannounced that it looks hawful. Which is wot Mum always says about the previously ginger one's hair-dying attempts, and so he fort he was being a supporterive husband and presenting a United Parental Response.

A United Parental Response is when Dads have to say wot Mums tell them to say; it is wot 'they' fink and wot 'they' will say when 'they' is asked.

But, for some strange and currently hunfathomable reason, Mum finks the previously ginger one's hair is fabumazing, and matches-her-skin-tones and looks really very well done, so Dad has binned quite actual scowled at and told not to be so rude.

Dad is feeling very hard dunned by now. He was only trying to be a loyal husband, and really doesn't give a stuff wot colour the previously ginger one chooses to have her hair, so long as Mum doesn't moan about it. So wotever it is Mum finks about the pink hair is wot he finks, though he'd rather not fink about pink hair a-very-tall, if that's okay with everyboddedy, and he promises never, hever to offer a hopinion again.

December 4

Has you ever had one of them days where you just do know that peoples are not going to believe you when you say wot happened, even if they does know you is a luffly boykin who always does tellering the truth?

Today was One. Of. Them. Days.

It started when the previously ginger one decided she needed to very hadmire the back of her head by standing on the stairs and looking in the mirror in the hall. And then she falled down the stairs. Backwards. You has never, ever heard a noise like it. Unless, of course, you has heard the sound of somebodddedy yellering and screaming as they falled backwards, taking every single banister pole with them, missing the big, fick, heavy post at the bottom of the stairs with their head, but bashing it with their hip, and then landing with a fuge FUD in the hall on top of a pile of maggy-zines and a dog bed.

And then having the banister rail land on their belly.

If you has hearded that sound then you will know that wot comes next is silence as everyboddedy else in the house waits a tiny second for SIGNS of LIFE from the person who has just falled down the stairs. They don't mean to wait but they do, cos they would prefer to know if they is going to be running towards a dead boddedy or somefing they can actual help with.

Fortunately, the previously ginger one did offer a SIGN of LIFE. Hunfortunately, it is not somefing I can repeat.

At this point, Dad hobbled in from the shed, which was a good blinking job because Mum was hinsisting that the previously ginger one did not move, and the previously ginger one was yelling that she was perfickly fine, or would be if Mum would stop holdering her down, get out of the blinking way and let her stand up. Dad made the hexecutive decision that noboddedy could be making that much noise and have a broken neck, so she could do standing up.

Then the phone rang. Gran-the-Dog-Hexpert wanted to know if Mum could get to Granny Mary's house as soon as very possibibble because Granny Mary had falled over and needed someboddedy to sit with her until the fuge green men with flashing lights could get there with their hambulance. And Mum was closest.

Mum had to decline Gran's kind hinvitation as she was about to take the previously ginger one to Opital, cos despite all the stairs clapsing round her and very actual breaking her fall, she still needed to get her head checked, even if the previously ginger one kepted saying she was FINE. Happarently, people's who has just dunned falling down the stairs and could have a head hinjury, and Dad's wot have got a broken foot, and have taken so many painkillers he doesn't know wot he's talking about, don't get to do choosering whether or not they go to Opital.

And that is how, this evening, four generations of my famberly ended up in the same Opital Hemergency department at the same time. This is not somefing wot I would want to hinflict on anyboddedy who isn't related to them. And hespecially as Granny Mary had to wait for nearly three hours for a hambulance to turn up, so that when she heventually arrived, the Opital had to actual cope with Gran-the-Dog-Hexpert and Mum making a formal complaint about the hambulance wait; the previously ginger one roaring that she wanted to go home cos there was nuffink wrong with her head, and Granny Mary furiously trying to stop her hentire famberly MAKING a SCENE, and could they all SHUSH! and stop hembarrassing her or she would disown them. Most of all Mum and Gran-the-Dog-Hexpert, who both looked like they'd spent a month living in a dog kennel, but apart from the previously ginger one whose hair looked pretty.

I knew you wouldn't do believering me ...

December 5

According to the previously ginger one, everyfing-hurts-I-can't-even-wear-my-glasses-to-See. My. Phone. wot is apparently the worstest fing that can happen to a teenager. Even her hair hurts, she says. We is making no comments about that, cos if she hadn't been actual hadmiring her pink hair, none of the falling-down-the-stairs fing would've dunned happening.

Mum fort about being furious with Dad about the banisters being wobbly, and landing on the previously ginger one, but then she realised that if Dad *had* dunned fixing them and nuffink had been wobbly, instead of them clapsing round the previously ginger one as she falled, they would have dunned killing her. But cos they all falled down, they saved her life. Mum is trying not to fink about that too much.

Dad says he isn't fixing the banisters in the house until the previously ginger one leaves home. Or stops dying her hair. Neither of them fings look like they'll be happening any time soon.

December 7

Me and Harry did exerlent running away from the hole stairs-clapsing-previously-ginger-one-flying hincident when it happened, and we both gotted over the hole noisy disaster quite actual quickerly. Going up and down the stairs with no banisters is not bothering us a-very-tall cos we is hagile and four-leg-drive doggies wot never used the banisters anyway. Dad is finding it most actual difficult, though, because of his broked toe, but is becoming most exerlent at bumpering down the stairs on his bum, or crawlering up on his knees, wot Harry finks is a signal for a game, so keeps trying to pull off Dad's socks if he takes too long.

Dad says he's going back to work tomorrow. It's got to be safer than being at home.

December 9

Tomorrow night is the Naward Ceremony, and noboddedy has got a clue wot to wear. I has got No Hidea why hoomans make so much fuss about wot to wear. I fink they should do wot I do and wear nuffink.

Nuffink is currently wot the previously ginger one claims she has to wear but this is Not. True. She has more actual clothes than everyboddedy else in this famberly put together; she just can't find hexactly wot she wants to wear, and is starting to fink about hinsistERing that Mum finds her somefing noo from the shops.

Hunfortunately, the previously ginger one is not Mum's pri-orry-tree: Dad is. Dad has got one suit for the summer, and somefing called a Dinner Jacket wot also has matching trousers, although, for some reason, usually nobbodedy ever talks about Dinner Trousers, just Dinner Jacket. The Dinner Trousers are getting a lot of talking about tonight, though, cos they is missing. Dad finks the fuge ginger boyman must have nicked them and taken them away to Universally. Whilst the fuge ginger boyman might have nicked the jacket, the shoes, and the weird windmill-type tie wot is all part of the Dinner Jacket, the trousers wouldn't actual fit him cos he is quite very slim and Dad is, erm, fifty-three.

But cos Dad can't wear the Dinner Jacket without the Dinner Trousers, Mum has got to actual guide him towards safe and sensibibble bits of his wardrobe, and persuade him to wear somefing that quite very actual matches. Vaguely. And convince him he can't wear a summer suit with a fuge great jumper hunderneath it to make it into a winter suit.

Mum's fort about asking her friends wot the mother of a Ward Nommy-knee should wear, but they will probababbly come up with all kinds of quite actual helpful but hunachievable ideas. Wot Mum really needs to know is wot the wife of a boat-builder who doesn't care wot he looks like should wear. And also whether the previously ginger one is still alive hunder the ignormous pile of rejected clothes in her bedroom.

December 10
Me and Harry are staying Home Alone tonight when Mum, Dad, and the previously ginger one go to the posh hotel in Norwich for the Naward Ceremony. We has binned Very Good Boys about not rubbing our hairy boddedies all over Mum's tights, and staying out of the way of the hairspray squirting going on at an halarming rate in the previously ginger one's bedroom.

There has binned a lot and a lot of talk about it being a wonderful fing in its Own Right to be nommy-nated, and noboddedy should be actual hexpecting the previously ginger one to win. And also to go very easy on the free wine. And not to cry if she doesn't win. Or if she actual does. Mum and Dad, that is, not the previously ginger one. She's got everyfing quite very actual together and horganised in her head. It's just her blinking parents who has turned into blithering hidiots needing to give themselves a pep talk.

December 11
Mum bawled her blinking hi-balls out, happarently, and Dad was driving so he couldn't have a drink of the free wine. But it turns out lots of people fort that being brave-in-the-head was just as actual good at being brave-in-the-boddedy, cos the previously ginger one did win the Ward for Houtstanding Bravery.

Mum had just a-very-bout stopped crying, and Dad started to ask if the previously ginger one would like a glass of wine to celly-brate, when there was a hannouncement that the Overall Star of Norfolk and Waveney was ... the previously ginger one! And she'd one the Hole Blinking Fing. So Mum started crying again and Dad had to snatch the glass of wine out of the previously ginger one's hand, cos she had to go up onto the stage to collect the Big Ward. And be hinterviewed, wot happarently she did cope with very quite well.

The previously ginger one can't remember a single word she said, though, but that's very okay because there was so many peoples there from the noospapers and the radio stations that we are sure writted it all down. Dad says it will be on the front page of the local papers tomorrow. By then, Mum will have stopped crying ... hopefully ...

December 12
Harry founded an I-Scream tub in the previously ginger one's bedroom. Mum very quite kindly did say that he could have it after quite a-very-bit of muttering about peoples-who-don't-know-how-to-take-fings-downstairs. But she didn't notice that it still had some I-Scream in it, and it wasn't as hempty as she fort.

Harry would like everyboddedy to know that, unlike actual me, he does love I-Scream quite a-very-lot and a lot. He really, wheely likes shaking the pot about so that melted strawberry I-Scream splatters every-blinking-where, and so he has some hinteresting and hunexpected treats to sniff out and lick off the walls later on today.

Mum would like everyboddedy to know that she does not like I-Scream any more. And the previously ginger one might be a Ward-Winning Star, but currently she is in the doghouse as-very-well.

December 14

We got a Crispmas Card today with a picture of some luffly boykin Collies wearing Father Crispmas hats. And there was a cat in the picture as well. Before anyboddedy starts gettering any stoopid and oppy-mistic ideas round here ...

DRESSING UP FOR CRISPMAS

I are *not* going to wear a Crispmas hat
I are *not* doing sit with Gandhi the cat
There's no way I'm posing in the snow
Forget all that cheese, the answer is no
I are *not* dressing up as a Crispmas elf
If you want an elf you can be one yourself

I are *not* not lying down by the Crispmas tree
With tinsel and baubles: it's just not me
I are *not* sporting antlers to look like a deer
I fink I has made that perfickly clear
I'm off to my bed with a run and a hop
If you want a Crispmas picture, use Photoshop

In my quite very own defence, I would like it to be actual known that, so far this year, Mum has tooked over 11,000 photos of actual me. And about a fousand of Harry.

We has both binned very quite tolly-rant of the clicky-whizzy fing, heven when Mum got a hupgrade on her phone, and discovered the burst hoption, so she can push down her finger and it takes ten photos in every single second.

But me and Harry will not be wearing Crispmas hats and posing for photos, fanking oo kindly. We does decline, no matter how many bitsa cheese there is on offer.

December 16

Today, we did go to the Harbour Cafe for a cuppatea after some himportant work belting-about-on-the-beach. At the cafe we did meet some quite very ickle dogs called Rolo and Fudge. Harry was actual confuddled about the ickle dogs, so I did have to give him some advices about them.

****HOW TO MAKE FRIENDS WITH ICKLE DOGS****

♥ The first very actual fing to do when you want to make friends with ickle dogs is to be a luffly boykin to their Mum. So then she will not fink you is a scary monster who is going to squish their little friend, and do relaxing and letting us doggies get on with saying hello

♥ Ickle doggies do still want to get to know you by having a good sniff. Even if that does mean they has to stand on tippytoes, and make you actual jump cos you losted sight of where they actual are. And then discover they is hunderneath you

♥ And they don't like being bonked on the head with sandy paws any more than big dogs

- ♥ Some peoples fink ickle dogs do bark a lot. That is cos they can't do running away if things get scary. Well, they could, but barkering is much, much quickerer if your legs is honly four inches long
- ♥ Benches is useful when meeting ickle dogs. You can have a meeting of minds. And also not do sitting on the ickle dog haccidentally
- ♥ If there is no benches for the ickle dog to stand on, fuge boykins should do lying down so that all the unequal leg fing goes away, and everyboddedy can be on the same planet
- ♥ Small dogs do sometimes have cute tricks that they can do to get bitsa cheese scone. Do not try to copy their sit-up-and-beg trick. You is a Lurcher and you don't bend that way. You will do falling over and makering a plonker of your actual self
- ♥ Small dogs do have to learn these tricks cos they is too small to counter-surf. Or jump on the table
- ♥ Do not do gettering off your soft blankie or Mum's coat or whatever fing you has found to lie down on cos you will lose it to a small dog. And they can be quite very actual stubborn about giving it back
- ♥ Most ickle dogs are not precious babies. Heven if sometimes they is stucked with a Mum wot does fink that way, and they get dressed up in all kinds of not-dog-but-dolly stuff

Rolo and Fudge were fabumazing, and didn't have to wear pink sparkly collars or other Orrendous dressing up stuff, wot I do fink is quite actual lucky.

I fink Harry did quite very well at Making Friends with Ickle Dogs, and I fink he does hunderstand that they is just like fuge boykins. Only shorter. And betterer at getting bitsa cheese scone.

December 17
Mum went to visit a good actual friend today, and she did take Harry with her and Not Me. You might be finking that I would feel a bit quite very Hoffended about this, and do Complaints to the Management, but I was actual pleased with this harrangement because I did get some Hempty Space.

Hempty Space is very actual himportant and hessential for a luffly boykin like Worzel Wooface. I do try to be tolly-rant and a good host, but sometimes it does all get quite actual too much, and then I does not want to play, and I does not want to be teaching Harry his ropes, and mostly I would like to be lefted alone to have some finking and snoozing time, without Harry dropping somefing hexciting on my head. Or trying to get into the Same. Bed. As. Me.

So I did get that time today, and it was quite very actual exerlent. Harry did have a fabumazing time visiting, and he did meet some strange cats. Well, all cats are strange, of course, but he did not know these strange cats a-very-tall, and Mum got to see how he got on with these new ones. He did quite actual well, according to Mum, and did show off his Sidekick-Hability-of-Finding-Shoes, wot Harry does fink is another one of his special talents.

Mum wishes he didn't have this special talent. And also that he would drop his Climbing-on-the-Kitchen-Table trick, but Dad reckons Harry is stuck with these talents whether or not we like it.

December 18
The fuge ginger boyman is coming home for Crispmas, and Mum says we

have got to Tidy. Up. Seriously, put everyfing away and get it all clean and tidy. In case you are actual finking that Mum is going to a fuge lot of troubles to welcome home the fuge ginger boyman, and everyfing has got to be clean and booful so he feels special and himportant, I does fink I should probababbly tell you the real reason Mum is tidying up.

The fuge ginger boyman is the single, most fugely huntrainable, messiest lump of hooman any of us has ever metted. There isn't a room he cannot turn into a muddle within about ten actual minutes. And Mum's feery is that if everyfing is spotless to start off with, and there is less stuff lying about for him to muddle, the longer it will be before he loses somefing and creates an Hapocolips lookering for it, or decides that somefing looks hinteresting and gets it all out and doesn't put it away.

At least that's the plan.

December 19
****FUGE NEWS****
Harry's Dad-wot-luffs-him has dunned sorting out everyfing! He has been working at night and sleeping in the car during the day so he can save all his moneys, and now he has got a-very-nuff to be able to rent a house. And wot is even betterer, he will be able to have Harry with him at the house! He has got a job wot Harry can go along to, and when he can't, there is a dog-walker who is going to look after Harry during the day. And does that all sound quite very actual okay?

Mum said this did sound very okay hindeed, so it has binned arranged that Harry will be going back to his Dad-wot-luffs-him after Crispmas when everyfing is settled and calm, and there will be time for them to get to know each actual other again before Harry's Dad has to start work.

I fink I are mainly quite actual pleased about this. Harry is a super boykin, and I are hoping that I will get to see him again once he leaves here, but I do hunderstand that it isn't always possibibble. Cos people's lives move on. And the hole point of fostering doggies is so that they can go onto the next bit of their lives, and another doggy can come and be saved or helped to feel betterer.

Harry hasn't needed to be saved or feel betterer, though: his Dad did just need a very little bit of time. And I fink we has mainly binned good for Harry. He can cope with cats now, and just about sleep in his own bed if he has to. But he still jumps onto the kitchen table, so at least his Dad will not fink Harry has becomed a different dog.

We has all decided that we shall actual henjoy our last week with Harry, though I fink we shall all do missing him a lot and a very lot.

December 20
Harry has learned how to do 'down.' I aren't speaking to Harry currently. I do fink he is seriously lettering the hole side down.

And making me look stoopid.

December 21

On the other actual paw ... Harry can't be left hunsupervised around cushions without them hexploding. Wot I do fink is not out-very-dunned by 'down.'

December 23

Do you know where the fuge ginger boyman's wallet is? He's honly binned home from Universally for a day and he's already losted it. I don't where it is but, according to the fuge ginger boyman, everyboddedy must have hidded it or pinched it or put it somewhere-very-actual-deliberately where he cannot find it.

I does wish he would stop yellering about it, but mostly I does wish he would blinking find it and stop stompering around the place whining about not being able to go to the pub. And also if he stopped doing all the lookering with his mouth and used his hiballs, he might find it.

I does luff the fuge ginger boyman quite very muchly but I doesn't fink that Universally he's at is very much actual cop. He's binned at that Universally for over two years now, and they hasn't taught him nuffink useful as far as I can see. He says he can clone all kinds of hinteresting and dangerous fings, but he still can't keep any of his fings safe. And he can't clone his wallet wot would be quite actual useful, and maybe then he'd stop sulking and saying that Christmas. Is. Cancelled cos he can't buy a pint.

December 25

As far as I are concerned if it comes to a choice between Crispmas and my routine, I would prefer to have my routine, fanking oo kindly. And the first part of my routine each day is that Mum gets out of bed to make a cuppatea, and I himmediately leap into the warm bit of the bed she has lefted.

Honly today, cos it is Crispmas, that space gotted hinvaded by a quite actual cheeky previously ginger one, who wanted to open her stocking with Mum and Dad. And then the fuge ginger boyman decided he wanted some space there as-very-well.

I would like to point out that the fuge ginger boyman is twenty-two now, and the previously ginger one is nineteen. They should both be feeling quite actual growed up and boring by now, and not want to do famberly time on Crispmas Day on my bit of the bed.

Mum says I are quite very actual right: they should both be too very growed up for this kind of actual fing. But she is quite actual glad they is not. And I should budge up.

December 26

I got a noo Oinky Pig for Crispmas from the fuge ginger boyman, and I has binned a luffly boykin sharing it very quite nicely with Harry. We has binned oinking away all day. Dad keeps chuckering it into my bit of the garden, but Harry is exerlent at finding it and bringing it back in so we can carry on with our oinking games. And yipping at each other and chasing the oinky pig up and down the hall.

I fink Dad is quite actual pleased that the fuge ginger boyman remembered how much I loved my last oinky pig wot got losted hunder suspicious actual circumstances, and never found, cos he keeps saying fank oo SO much to the fuge ginger boyman. And chucking it out in the garden again.

December 27

Tomorrow, Harry is going home. Back to his Dad-wot-luffs-him. We is so very actual pleased for Harry, and very, very quite proud of Harry's Dad-wot-luffs-him, and we does hope they have the most actual fabumazing life together, forever again.

Harry's Dad says he has a fuge bed, a really, wheely fuge, ignormous plenty-of-room-for-Harry bed. Harry won't have me to keep him comp-knee when he goes home, so he has decided that Harry will sleep on the bed with him. I don't fink Harry will miss me very much now.

Everyboddedy here does have mixed forts. We will miss most of Harry. Not the standing-on-the-table-and-shoutering-at-the-Hoover bits, but halmost everyfing else. I will miss our playtimes and bitey-facey, and Mum will miss his cuddles and his clever hobedience bits, and Dad will miss Harry's very fusey-tastic welcome homes (wot I aren't as good at, to be quite actual honest).

Harry did always have a Dad-wot-luffs-him, and Mum was quite actual bossy to him about making sure he came back for Harry, cos Harry wasn't our dog.

So, everyfing has turned out perfick. Just don't tell Harry he has got to have a bath tonight ...

December 31

Dear Harry

It's binned a few days since you did go home, and fings here has started to return to very actual normal.

I has missed you a ickle bit I has got to actual say, and dunned some wandering round lookering for you, but I has been to see Kite and had some playtimes, wot is sort of making up for it.

The cats would like you to know that they isn't missing you a-very-tall. And I aren't henjoying them not missing you, to be quite very honest. Now that you has gonned, they is popping up in all kinds of weird and hunusual places wot is not always funny. Or good for my hiballs.

I does hope you and your Dad-wot-luffs-you are gettering on actual okay, and learning all about each other again. And that you is being a exerlent boykin.

From your luffly boykin

Worzel Wooface

Pee-Ess

It does seem that my oinky pig has gonned. Did you take my new oinky pig home with actual you, or has Dad tooked fuge hadvantage and decided to Throw. It. Away. when I was too busy wondering which bit of the big bed was the warmest? If you does have it, could you actual let me know cos I are currently sulking, and I does not know who I should be sulking at the most: You, or Dad.

Heppy-log

Dad says a Heppy-log is useful when somefing happens after the end of the book wot can't wait for the next one.

For a long actual time we was quite very hopeful that the hole saga of the beach ban would be sorted out before it got to Noo Nears Eve, but Cow Sells have this really, wheely complercated way of doing fings wot takes for-HEVER. And then a meeting got postponed, and it was dragging on for so long I was starting to fink it was never going to get sorted out. Then, All. Of. A. Sudden. there was noos. I has decided it is my himportant work not to do dragging fings out any-blinking-longerer, so I are going to tell you actual now ...

WE SAVED THE BEACH!

The hole Southwold Beach ban has binned sorted out, and the Cow Sell and the Cow Sellers and fabumazing Auntie Charlotte and SARDOG has reached a com-pree-wise, which is when hoomans do start listening to each other's actual words and forts, and do sorting fings out until noboddedy does feel hard dunned by. Or until noboddedy has gotted their own actual way and they is all cross, Dad says.

Us doggies love the beach, hespecially in winter when noboddedy else actual does. And it is quite very himportant that we can have running and playing times. For lots of actual doggies, the beach is the honly safe place that they can do this fing, and then we do go home and zonk out in front of the telly and are luffly boykins. Doggies wot don't have a-very-nuff hexercise get bored and do eating cushions. Or chasing cats. Or worser fings. Everyboddedy does keep going on and on about Dog Safety and Responsibibble Dog Ownership, but Mum reckons it is blinking hard to be responsibibble if there is nowhere to do taking your dog for a walk. And it gets hexpensive in the sofa department as well.

But – and this is the himportant bit – the Cow Sell will be lookering again at the hole Dogs-on-the-Beach fing on a Nannual Basis, and if peoples don't pick up after their doggies, or there are complaints about people lettering their dogs on the wrong bit of beach in the summer, or dogs are left to wander about and not being watched, then the Cow Sell could do banning dogs all over again ...

So, if you does come to Southwold. Please. Please, Quite-Very-Actual Please, follow the hinstructions on the booful noo SARDOG signs about where you can be. And help to keep the beach safe and clean, and most himportantly dog-friendly for everyboddedy.

THE END
(for now)

HELP US TO KEEP SOUTHWOLD A DOG FRIENDLY TOWN

DOGS MUST BE **KEPT ON A LEAD** ON THE PROMENADE A L L Y E A R R O U N D

DOGS ARE **NOT ALLOWED** ON THE BEACH IN FRONT OF THE PROMENADE BETWEEN 1st APR AND 30th SEP

THERE ARE **NO RESTRICTIONS** ON THE BEACH BETWEEN 1st OCT & 31st MAR

ALWAYS **PICK UP** AFTER YOUR DOG and KEEP YOUR **DOG UNDER CONTROL** NOT EVERYONE WANTS TO PLAY!

Thanks for the graphics **Mike&Scrabble** www.mikeandscrabble.com

Southwold Town Council
http://www.southwoldtown.com/

Waveney
District Council
www.eastsuffolk.gov.uk

Southwold & Reydon Dog Owners Group
www.sardog.co.uk

156

HELP US TO KEEP OUR
COMMUNITY
DOG FRIENDLY
CHILDREN
PLAY HERE

ALWAYS **PICK UP**
AFTER YOUR DOG

DO **KEEP YOUR DOG** ON A **LEAD**
AND **STICK TO THE PATH**

Thanks for the graphics **Mike&Scrabble** www.mikeandscrabble.com

Southwold Town Council
http://www.southwoldtown.com/

Southwold & Reydon Dog Owners Group
www.sardog.co.uk

"Beautifully written, extremely witty & heart-warming – literally couldn't put it down – loved it!"
Dog Training Weekly

"I've actually met Worzel Wooface. I asked him what he thought of this book. From the look on his face, I guessed that he was quietly pleased, and so he should be"
Paul Heiney

"A real feel-good, laugh-out-loud book ... a must for all Lurcher owners"
Daily Express

***** AMINE REVIEWS
✶✶✶✶✶ **AMAZON REVIEWS**
• "A very actual fabumazing book!"
• "Bestest dog author ever is back!"
• "Worzel back, funnier and wiser"
• "Fabumazingly actual awesome!"
• "Am already looking forward to the next one!"

"Woo 2: required reading for anyone who has ever given their heart to a dog to tear"
Geelong Obedience Dog Club

The first book in a fabulous brand new series for children from Worzel Wooface!

For children of all ages!

Children and dogs can be the best of friends, and good for each other in so many ways.

Sadly, though, dogs are often misunderstood by both adults and children, and their behaviour misread.

Worzel Says Hello! is the first in a new series of delightful educational and fabulously illustrated guides to understanding how dogs think, feel and behave, so that all children can have a wonderful relationship with the dogs in their lives, and all dogs can feel happy, safe, and loved.

HH5160 • Hardback • 20.5x20.5cm • £6.99* • 32 pages • 49 colour pictures, inc 45 original illustrations • ISBN: 978-1-787111-60-8 • UPC: 6-36847-01160-4

For more info on Hubble and Hattie books please visit www.hubbleandhattie.com; email info@hubbleandhattie.com; tel 44 (0) 1305 260068
*prices subject to change

Catherine and Worzel

Catherine Pickles is a full-time family carer, writer and blogger. Her blog about Worzel reached the final of the UK Blog Awards in 2015.

Worzel Wooface is a Hounds First Sighthound Rescue dog who likes walking, spending time with his family, and chasing crows when given the opportunity. His current hobby is chewing wellies on unmade beds. He lives in Suffolk.

Catherine has fostered numerous sighthounds for Hounds First Sighthound rescue. Her hobbies include sailing, walking, gardening and amateur dramatics, most of which she likes to do with Worzel (apart from gardening, but she doesn't have much choice about it, and the amateur dramatics, which he would hate). She, too, lives in Suffolk, with her husband, two nearly grown up children, and five cats.